# PLATO'S ALARM CLOCK

# PLATO'S ALARM CLOCK

## AND OTHER AMAZING ANCIENT INVENTIONS

JAMES M. RUSSELL

MICHAEL O'MARA BOOKS LIMITED

First published in Great Britain in 2018 by
Michael O'Mara Books Limited
9 Lion Yard
Tremadoc Road
London SW4 7NQ

A CIP catalogue record for this book is available from the British Library.

Papers used by Michael O'Mara Books Limited are natural, recyclable
products made from wood grown in sustainable forests. The
manufacturing processes conform to the environmental regulations of the
country of origin.

ISBN: 978-1-78243-934-9 in hardback print format
ISBN: 978-1-78243-935-6 in ebook format

1 2 3 4 5 6 7 8 9 10

Designed and typeset by D23, London

Printed and bound by CPI Group (UK) Ltd, Croydon, CR0 4YY

www.mombooks.com

# CONTENTS

## CHAPTER 1
## EVERYDAY LIFE

## CHAPTER 2
# MECHANICAL AND INDUSTRIAL TECHNOLOGY

## CHAPTER 3
# MYSTERIES OF THE ANCIENTS

## CHAPTER 4
# MILITARY INVENTIONS

## CHAPTER 5
# MEDICAL KNOWLEDGE

## CHAPTER 6
# SCIENTIFIC ADVANCES

# INTRODUCTION

In the modern world, we like to think we are pretty smart. We have the internet, amazing medical knowledge, space flight and self-driving cars to prove it. However, if we were stranded on a desert island tomorrow, most of us would have no clue how to start a fire or catch a fish, let alone rebuild all that extraordinary technology we rely on.

We're not necessarily any cleverer than our ancestors – we just have an accumulation of centuries of technological progress on which we can rely. Many of the ancients were much more advanced than we realize. There are recent inventions that had actually been discovered centuries earlier and then forgotten. A huge proportion of our modern devices and machines rely on ancient inventions such as paper, levers and gears, and many everyday items are far older than we realize. The Aztecs had chewing gum. There was brain surgery in the Stone Age. And the Chinese have had silk clothes, sulphur matches, toilet paper and whiskey for thousands of years.

That's not to mention those pieces of ancient technology which we no longer understand, or which surpass our present-day knowledge. We don't know how to make Damascus steel, which was once the hardest metal in the world. We don't know how the Mayans made their weatherproof pigments. And we don't know the secret of Greek fire, a much-feared mysterious Byzantine chemical weapon that burned in water.

This book collects the stories of hundreds of ancient devices, inventions and breakthroughs from around the world and across the centuries. From the first telescope to the invention of the crossbow, and from Etruscan false teeth to early Chinese mechanical clocks, this is a fascinating glimpse into past eras that were far more technologically complex and advanced than we sometimes realize.

## CHAPTER 1

# EVERYDAY LIFE

# The Calendar

## First Invented: Scotland
## Date: 8000 BC

From earliest history, mankind has been fascinated by the sky and its two most prominent occupants – the sun and the moon. It was only natural that we should start to keep track of the fluctuations in these two celestial objects. The passage from day to night is one obvious interval, as is the cycle from new moon to full moon and back again. And for hunter-gatherers or early farmers, it was also crucial to understand the rotation of the seasons through the year.

Making sense of these various measures was a complex task. For instance, there is no tidy number of lunar months in a solar year. Twelve lunar months take about 354 days and you either need to find a way to add those odd days back into your system or to accept a gradual slippage in how the years relate to the seasons. And that's before you even start to think about the problem of leap years . . .

Calendars of various sorts were clearly in use before the Bronze Age. We have written records of calendar systems from the Sumerian, Egyptian and Assyrian civilizations dating to about 5,000 years ago. However, one recent archaeological find in a field in Scotland, at Craithes Castle in Aberdeenshire, suggests that calendars were in use much earlier than that.

The site contained a series of twelve pits, which seem not only to show the phases of the moon but also to monitor lunar months. The pits, which have been dated to 10,000 years ago, also align with the midwinter sunrise. This would allow the hunter-gatherers who created them to reset each year correctly to align with the seasons, which suggests significant levels of understanding and sophistication in the pre-agricultural Mesolithic people of the area. The academic Vince Gaffney, who was in charge of the scientific analysis of the site, said that it 'illustrates one important step towards the formal construction of time and therefore history itself'.

# Plato's Water Alarm Clock

## First Invented: Fourth century BC

We sometimes imagine the past as a time without clocks, when everything moved at a much gentler pace. Of course, life is not as simple as that, and there have been many historic situations in which people needed to find ways to keep to a busy schedule. For instance, the Greek philosopher Plato (427–347 BC) wanted a way to get himself and his students out of bed in time for lessons. As a result, he became the inventor of the alarm clock.

Simple water clocks – in which the gradual drip of water into or out of a vessel is used to record the passing of time – had been in existence in Babylon and Egypt by the sixteenth century BC. It is also possible that such clocks were used earlier in India and China, as long ago as 4000 BC. The innovation in Plato's water clock was that it also featured an alarm. A vessel was gradually filled with water, until it reached the height at which a tube led out of the first vessel into a lower receptacle. The tube functioned as a siphon, meaning that as soon as water started to drain out through it, the rest of the water was sucked into the tube with it. As a result, all of the water was instantaneously dumped into the lower receptacle. This lower receptacle was almost completely enclosed, except for a few small openings designed to act as whistles when air was forced through them – which happened when the water fell. So Plato's students were woken up, along with their teacher, by a loud whistling noise emanating from the extraordinary alarm clock.

Other early alarm clocks worked in similar ways. One involved a vessel that filled with water until it became heavy enough to fall and clatter onto a table below, making a loud noise in the process. Another design used a candle with a metal ball embedded in it, which burned down until the wax around the ball melted and fell onto a metal surface.

# Beekeeping

Mesolithic rock painting of a honey hunter.

Bees are older than humans in evolutionary terms, and throughout human history we have been fascinated by the problem of how to get at their honey. A Spanish rock painting from 8,500 years ago shows men stealing from the nests of wild bees. However, the history of beekeeping, where the bees are kept in artificial hives, is a much more recent story. A temple at Abu Ghorab in Egypt, dating from the Fifth Dynasty (2500–2400 BC), shows successive stages of honey manufacture – from taking honeycomb from the hive to draining the honey into jars. It was a key ingredient of many Egyptian medicines, as well as being used in cooking. The Egyptians made hives from dried mud, while the Greeks and the Romans refined the method using clay hives. The scale of manufacture in ancient Egypt is shown by the fact that in the twelfth century BC an offering of over 30,000 jars of honey was made to appease the gods. One of the few ancient civilizations to reject the use of honey was the Spartans, who described cakes made from honey in rather macho terms as 'no food for free men'. The more enthusiastic attitude taken by other cultures was summed up by a Roman blessing: 'May honey drip on you.'

# The Mechanical Clock

## First Invented: China
## Date: Eighth century

Sometimes the joy of history lies in the small details, such as the original names of inventions. The world's first mechanical clock went by the name of the 'Waterdriven Spherical Birds'-Eye-View Map of the Heavens'.

Invented by Yi Xing, a Buddhist mathematician and monk, in AD 725, it was developed as an astronomical instrument that incidentally also worked as a clock. In spite of the name it wasn't strictly speaking a water clock (one in which the quantity of water is used to directly measure time). However, it was water-powered – a stream of falling water drove a wheel through a full revolution in twenty-four hours. The internal mechanism was made of gold and bronze, and contained a network of wheels, hooks, pins, shafts, locks and rods. A bell chimed automatically on the hour, while a drumbeat marked each quarter-hour.

Another splendidly named clock was the 'Cosmic Engine' built by the Chinese inventor Su Song between AD 1086 and 1092 for an emperor of the Sung Dynasty. This was also a mechanical astronomical clock, but it was huge, spreading over several storeys in a tower that was over 10 metres (35 feet) high. It was made of bronze and powered by water. At the top, a sphere on a platform kept track of the motion of the planets. The clock remained in place and working until 1126 when it was lost in a Tatar invasion.

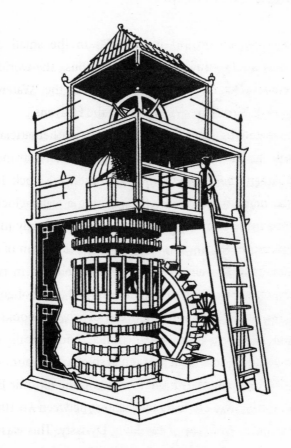

The Chinese engineer Su Song's
hydro-mechanical clock tower.

# Mesoamerican Chocolate

## First Used: Central America
## Date: c. 1750 BC

Today chocolate is a tasty treat enjoyed around the world, but did you know that it was once the food of gods, and that it has even been used as money? The cacao bean grows in the wild parts of Central America. The seeds grow inside a long sheath, inside of which a sweet pulp contains about thirty-five beans. They taste quite bitter, and it was possibly the sweeter pulp that people first ate or drank. However, when they were fermented, the beans could also be made into a drink. The earliest archaeological evidence of this method of consumption comes from a clay vessel, from about 1750 BC, which shows the residue of a chocolate beverage. At this stage the drink would have been bitter, as sugar wasn't grown in the Central American region before the arrival of the Europeans.

We don't know much about how these early chocolate drinks were prepared, but evidence from the later Mayan civilization may give us a clue about the older traditions. In a document from about the thirteenth or fourteenth century AD, cacao is identified as a sacred drink, which was associated with Kon the rain god in particular. The Mayans believed that cacao pods were made ripe by droplets of the blood of their gods. They prepared their version of the drink, which was also believed to grant virility to the men who drank it, by combining a paste made from the beans with water, cornmeal and chilli, then transferring the bitter brew from one cup to another to produce a foamy topping. (Women were traditionally banned from drinking this concoction for fear of the effect this virile drink might have on them.)

In the Aztec Empire – which ruled a large area in modern Mexico (north of the region of the Mayans) from the fourteenth century until the Spanish conquest started in the sixteenth century – cacao beans were a valuable commodity. Because the beans didn't grow in drier weather conditions, the Aztecs started to impose taxes on their subjects in the south, making it known that these taxes were to be paid in cacao beans. This led to the beans being used more widely as currency. The early Spanish invaders noted that their prisoners treated the bean with great reverence, and interrogated them to discover its properties. The drink was initially imported to Spain in its original bitter form. However, the Europeans learned to add a sweetener in the form of honey or sugar, and the modern form of chocolate was born.

# The Umbrella

The first umbrellas were sunshades or parasols – this is partly because the most advanced civilizations developed in warmer climates. The earliest evidence we have of such umbrellas comes from 2400 BC. In a victory monument to Sargon, king of Akkad (in modern-day Iraq), he is depicted walking ahead of his troops while an attendant holds a parasol over his head to protect him from the sun.

By the first millennium BC, umbrellas had become a status symbol. The wealthiest Egyptians, for instance, looked down on suntans as being characteristic of the ordinary workers in the fields, and the pharaoh and other high-status individuals were often depicted with aides holding a sunshade over them.

The earliest parasols were fairly flimsy and not waterproof, so would have been useless in a rainstorm. For instance, parasols made in China from early in the first millennium BC were made of silk. For an all-purpose umbrella, we have to go forward to the Wei Dynasty (AD 386–533), when umbrellas started to be made of heavy mulberry paper that was oiled to make it resistant to water. From this point onwards the umbrella had a new function: it could protect its owner against a sudden downpour!

# A Brief History of the Lavatory

The dilemma of how and where people should 'go to the toilet' goes back to prehistory, when a hole in the ground would have been the most common solution. However, lavatories of one sort or another have been in existence for at least five millennia:

- In the Stone Age village of Skara Brae, on the Orkney Islands, archaeologists found what appear to be 5,000-year-old toilets in the stone walls of the houses, above drains that lead away from the buildings.

- From the same period, in the ancient city of Mohenjo-Daro in the Indus Valley in Pakistan, houses have been found with a similar arrangement: brick holes (with the remnants of wooden seats) over a chute leading to a communal drain.

- A palace at Eshnunna in ancient Mespotamia had a communal toilet, with six seats with raised brick seats in a row.

- In the seventeenth century BC, the Palace of Knossos on Crete featured a sophisticated system in which earthenware pans (in a room with easily cleaned gypsum slabs for walls) were connected to a water supply that flowed through terracotta pipes.

- An Egyptian toilet from about the fourteenth century BC was made from limestone with a keyhole-shaped opening, placed over a removable jar (which had to be cleaned out periodically). A hollow space on either side of the seat contained sand, to be thrown down the hole afterwards.

- The first known portable toilet also comes from this period in Egypt: a wooden stool with a large slot in the middle over a pottery vessel. It was found in a tomb at Thebes.

- The Roman Empire produced increasingly sophisticated toilets including, by AD 315, 144 public lavatories in Rome itself.

- The Romans also introduced the first-known hand-flushing toilets. However, their toilets were not always terribly hygienic – a surprising number were located in the kitchen, adjacent to food preparation areas. This was because the same arrangements were in place for kitchen and human refuse.

Marble seats in an ancient public toilet in Dion, Greece,
c. second century AD.

- Before toilet paper existed, moss was often used for
  the same purpose. Hay and straw were often left in
  piles to be used in the toilets of Western castles and
  monasteries.

# Beds and Bedding

## First Used: Unknown
## 5,000+ years ago

As well as containing one of the first examples of an ancient lavatory, the ruined village of Skara Brae on the Orkneys provides us with evidence that fairly complex beds were being used 5,000 years ago. This suggests that the idea was a much earlier one, which had been refined over time.

Wood is scarce in the Orkneys, so the inhabitants of the stone houses had used stone, their most plentiful resource, to create raised sleeping platforms. The stone bed-bases were set away from the external walls (which would have been the coldest) and adjacent to internal walls. The most likely material they would have used as a mattress is dried moss.

Beds and bedding continued to evolve over the centuries. By the medieval period, beds consisted of pieces of cloth or carpets stuffed with feathers, wool or human hair, to create a kind of mattress. Curtains or drapes would surround the sleeping area to shut out the light and to provide a measure of privacy (although in many poor households the beds would have been shared by several people). Wooden beds remained a luxury. From the twelfth century onwards we find examples of beautiful carved beds that would have been made for the wealthy. The first fold-up beds were also from the twelfth century – these functioned as a bed at night, but could be folded over into a couch during the day, just like modern sofa-beds.

# Locks and Keys

## First Used: Assyrian Empire
## Date: 1700 BC

Locks and keys are a fascinating historical development, as they represent both the growth of personal wealth and the need for privacy. The oldest lock that has been found comes from the ruins of Nineveh, which was the capital of ancient Assyria. It worked on the same principles as the locks of ancient Egypt, which were wooden pin locks (in which pins inside the lock are lifted into a recess by the key, removing the obstruction to opening the door).

In Homer's *Odyssey*, Penelope, the wife of Odysseus, is described using a key when she unlocks the door to a strongroom to fetch his bow. At this time locks were only used for very specific purposes, within wealthy or high status buildings. For protection from enemies a lock was insufficient – for instance, walled cities would be barricaded with more robust blockades. And there was no need to lock ordinary houses, as they contained little that was worth stealing. It was only within the most affluent and powerful buildings that there was a need to conceal objects or keep them safe in this way.

The Romans copied the technology of Egyptian locksmiths, and then improved on the techniques by using iron and creating the warded lock. Internal obstructions rather than pins were used to ensure that only the correct key would open the lock. Rich Romans often wore keys as rings on their fingers – this was a status symbol, as well as a way of keeping the key safe and in view at all times. It also demonstrates that the metalwork of the Roman period was sufficiently precise and advanced to produce locks and keys that were much smaller than those of the earlier civilizations.

# The Wig

The first wigs that we know about were in use in Egypt. The ancient Egyptians were obsessed with cleanliness. They washed several times a day, and saw being hairless as clean and civilized (as opposed to wild animals, or those they thought of as primitive peoples). Both men and women wore their hair cropped or shaved. Hair was removed by shaving, depilatory creams, or even by rubbing it off with a pumice stone. They wore wigs to keep the sun from burning their scalps. They were made of either natural or artificial hair, and were held in place by beeswax and resin. Wealthier individuals had elaborate decorative wigs topped with perfumed cones, which acted as a status symbol. Wigs were also in use at various times in other ancient civilizations, including the Assyrians, Jews of ancient Israel, Carthaginians, Greeks and Romans.

Egyptian canopic jar, c.1349–1336 BC.

# The Secrets of Distillation

## First Recorded: The Abbasid Caliphate/ Modern-Day Iraq
## Date: Ninth century AD

The two main methods of making alcohol are fermentation, in which the chemical breakdown of a substance such as fruit or grain produces alcohol; and distillation, in which a fermented drink is further purified through evaporation and condensation of various elements in a liquid. The oldest confirmed fermented alcoholic beverage to be identified comes from Jiahu in the Yellow River Valley of China, from about 7000–6600 BC. However, many scientists believe that fermented drinks would have been consumed in the Palaeolithic period 10,000 to 12,000 years ago. The 'drunken monkey hypothesis' goes even further. This is the suggestion that our distant primate ancestors discovered that they could get intoxicated by consuming rotting fruit from the forest floor, and that this knowledge persisted throughout the evolution of those primates into early hominids.

Distillation, however, is a much more recent discovery. There is some evidence from the first to third centuries AD that distillation was being practised in Roman Egypt and China. The Romans, for instance, extracted turpentine from pine oil. Many early sailors had improvised methods of distilling drinkable water from sea water, while some societies discovered ways of making reinforced alcohol through the freezing process – a fermented drink such as wine is cooled to the point where the water content freezes, leaving a much stronger residue behind.

However, the combination of a lack of evidence and the secretive ways of early alchemists and craftsmen means that we can't be sure whether or not alcohol was distilled in this period. So the first definitive fact we can be sure of is that the Arabian philosopher Al-Kindi was the first to

The early distillation process.

leave behind a written account of the secrets of distilling alcohol, giving a detailed description of the process, ending '… and so wine is distilled in wetness and it comes out like rosewater in colour'.

Al-Kindi lived in modern-day Iraq, in Basra and Baghdad, and is remembered as the father of Arabic philosophy. Al-Kindi's instructions helped to spread the secrets of the process far more widely, as different areas of the world learned to make 'fire-liquid', raki, whiskey, vodka, gin and much more besides. The word alcohol itself comes from *al-kohl*, because the process used by Arab chemists to make eye make-up was a similar process.

# Make-Up

## First Used: Ancient Egypt
## Date: 4000 BC

From a much earlier period than the time of the Arab chemists mentioned above, ancient Egyptian images are notable for the heavy make-up that is on display. From about 4000 BC onwards, the Egyptians used a variety of local ingredients to create cosmetics for magical, aesthetic and medical purposes. In particular they used the copper ore malachite to create green eye make-up, kohl (made from the sulphite mineral stibnite) or the cheaper charcoal for eyeliner, red ochre for rouge and lipstick, and henna for staining the skin. They applied make-up with sticks of wood, ivory or metal. The most common look was a combination of black lines to give the eyes a more pronounced look by giving them a more almond or feline shape, along with a wider area of green surrounding the eyes.

Part of the rationale for Egyptian make-up came from health and safety considerations. They believed that kohl repelled flies and warded off infections, while also keeping the wearer safe from the 'evil eye'. There was actually an element of truth in the health aspect of this – the kohl helped the skin to produce nitric oxide, which in turn strengthened its immunity to infection, while the sooty substance reduced the effect of the glaring sun on their eyes.

When it came to the lips, the ochre was often combined with resin or gum so that it would last longer. Cleopatra's personal cosmetic routine included a bright red lipstick made of flowers, crushed ants and fish scales, carmine and beeswax. As a result of her choice of red, this became the most popular colour among her subjects, but other popular colours for lipstick included orange, purple and a dark blue.

# The Fire Brigade

The first mechanical firefighting equipment was developed in Greece in the third century BC by Ctesibius and Hero of Alexandria (see page 64). Ctesibius, who is remembered as the father of pneumatics, invented a powerful water pump, and Hero adapted this into the world's first known fire engine, which used a combination of pistons, cylinders and valves to direct water onto a blaze.

However, for the first fire brigade we have to go forward to the first century BC in Rome, when a rich but unscrupulous individual called Marcus Licinius Crassus spotted a gap in the market and created a squad of 500 men. At the first sign of a fire, the brigade would rush to the burning building. However, they would then stand and haggle with the building's owner over the price for their services before even attempting to extinguish the flames. Unsurprisingly, Rome continued to suffer from many fires – including a serious one in AD 64, which burned two-thirds of the city to the ground.

# Bone Tools

## First Used: Africa or Europe
## Date: *c.* 50,000 years ago

The debate over when bone tools were first used throws up one of the more interesting quandaries of the early history of humankind, as it relates to the relationship between *Homo sapiens* and their cousins the Neanderthals. In Africa, there is evidence from 70,000–100,000 years ago of bones that that had been intentionally cut, polished or carved into pointed tools, awls, harpoons or wedges. However there is not much consistency in the types and locations of these archaeological finds.

In that period, the predecessors of *Homo sapiens* were still evolving in Africa (they didn't migrate to Europe until about 44,000 years ago). However, bone tools have also been discovered in Europe that date to at least 50,000 years ago, which suggests they must have been made by Neanderthals (who were already in Europe, having evolved there, about 200,000 years ago). So, while it may be that the two types of hominid discovered how to produce tools independently, scientists have also raised the intriguing possibility that those of our *Homo sapiens* ancestors who first migrated to Europe from Africa learned how to work bone from their Neanderthal cousins.

Stone tools were in existence much earlier than bone versions, but stone is cruder and less adaptable – the breakthrough came when humans learned to use stone tools to cut, grind and polish fragments of bone.

Tools can be made out of various types of bone, although antlers and longer bones are the most useful for creating long, sharp tools. Other early types of tool that have been found include fish hooks, cutlery, needles and pins, jewellery, and tools for scraping and polishing. Teeth were often strung together as jewellery, while animals' hooves could also be turned into a rudimentary musical instrument such as a bell or, if filled with small stones, a rattle.

# Cutlery

## First Invented: Worldwide
## Date: 100,000–12,000 BC

While stone knives were being used half a million years ago in Africa, humankind ate with their hands for most of prehistory. The first pieces of cutlery developed when people started to cover the handles of a stone knife with wood or skin in order to make it more comfortable to hold.

It's impossible to say when this practice started, but from the start of the Neolithic period (which was about 12,000 years ago) we know that such knives were in use, along with simple spoons made of hollowed-out wood, seashells or flint. And by 3,000 years ago, it was commonplace for people to use a combination of knives and spoons to eat their meals. At around the same time, the chopstick came into use in China.

Throughout the Bronze and Iron Ages, cutlery gradually became more sophisticated. However, forks had still not made an appearance at the dinner table. Fork-shaped tools such as pitchforks were used for other purposes but – possibly due to the expense and difficulty of creating a metal fork – it was not until the first millennium AD that they came into use as cutlery. The practice started with the use of ceremonial forks and 'flesh-forks' to remove meat from cooking pots in ancient Greece and Egypt. As an individual eating tool, the fork became popular in about AD 400 in the Middle East.

From there the utensil spread only slowly. When the Italian nobleman Domenico Selvo married a Greek princess in the eleventh century, she scandalized his compatriots by bringing with her the first forks to be seen in central Europe. For Christians of this period, forks were problematic – when earlier deities were co-opted into the Christian worldview, the trident of the Greek god Poseidon gradually transmuted into the pitchfork that was used by Satan to torment sinners. As a result, the princess's use of the fork was seen as heretical and deeply shocking.

Using a fork to transfer food into the mouth continued to be seen as uncouth, at best, for some time, but gradually the opposition to forks lessened, and they became accepted as members of the European cutlery family.

# Refrigeration

## First Invented: Modern-day Syria
## Date: 1700 BC

People were using snow and ice to cool their drinks and food long before the invention of refrigerators. By the time of the Roman Empire, snow was brought on sledges from the nearest mountains, then kept in snow pits or icehouses and covered in straw. As the top layer of snow melted it sank downwards and refroze into ice, while the bottom of the packed snow became increasingly impacted by the weight of snow above it. The import of snow and ice had reached such heights in the third century AD that the Emperor Elegabalus (who reigned from 218–222) was able to construct a miniature mountain of ice in his garden in the hope it would provide a form of air conditioning for his villa. The Romans and Greeks may have learned this art from Alexander the Great, who was described by the classical food writer Athenaeus as having dug 'thirty refrigerating pits which he filled with snow and covered with oak boughs'.

The Chinese knew how to keep snow and ice in icehouses even earlier. During the Tang Dynasty (618–907 AD), a text called the *Food Canons* describes the various rituals needed in the maintenance of icehouses, including the task of cleaning them to prepare for a new ice harvest. The icehouses would then be used for the preservation of fruit and vegetables. As early as the Chou emperors (fourth–third century BC), the court had an 'ice service' in which nearly a hundred staff worked, chilling food, wine and even corpses.

For history's first-ever mention of refrigeration, however, we have to go back to the Near East. In an inscription for Zimri-Lin, the ruler of Mari (an ancient kingdom near the border of modern-day Iraq and Syria), it is claimed that he had constructed an icehouse near his palace 'which never before has any king built on the bank of the Euphrates'.

# Razors

The oldest razor ever found has been dated to about 20,000 years ago. The first razors were made from clam shells, shark teeth, and sharpened flint. They became gradually more common through the Bronze Age, when bronze or obsidian (volcanic glass) was used to make sharper implements. Such objects were personal to the owner and would have been seen as status symbols – for instance, decorated gold and copper razors have been recovered from Egyptian tombs from about 4000 BC, indicating that they were of sufficient value to be buried with their owners.

Egyptian razor and mirror, c.1492-1473 BC.

# Chewing Gum

## First Used: Finland
## Date: 4000 BC

When the Spanish conquistadores reached Central America, they were intrigued by the gum-chewing of the Aztec prostitutes (who wore yellow cream on their faces, red cochineal dye on their teeth, and plied their trade on street corners). What they were actually chewing was chicle, a milky liquid sap that oozes from cuts in the wild sapodilla tree (*Manilkara zapota*), and which turns into a harder gum as it dries.

Chicle had been a sacred substance for the Mayan civilization centuries earlier. Their heroic snake deity Kulkulan (also known as the 'Feathered Serpent') was depicted chewing it in folk tales. However, after the Spanish subjugated the Aztecs, the trade routes from the forests to the cities collapsed and chicle chewing was only preserved in the forest areas.

The chewing of tree resin was also practised in other parts of the world. In North America, Europeans picked up the habit of chewing spruce resin from the Native American peoples, and this was the inspiration for the first commercially produced chewing gums. In 1848, John B. Curtis created 'The State of Maine Pure Spruce Gum', and in 1850 another gum based on paraffin wax sweetened with powdered sugar came on to the market. Further patents and adaptations soon followed – including a flavoured chewing gum that went on sale in the 1860s (manufactured by John Colgan, a Kentucky pharmacist), which was derived from the same chicle chewed centuries earlier by the Aztecs and Mayans. One of the minor ironies of history is that it was surviving Mayan Indians, hunting for sapodilla trees in the jungle because of the growing demand for gum in the United States, who stumbled on the now-famous ruins of many of the ancient Mayan cities.

However, for the origins of chewing gum, we have to delve much further back through time. The habit of chewing naturally occurring substances (for antiseptic, medicinal or recreational purposes) seems to go back to the Neolithic period, and to have sprung up independently in many different cultures. The ancient Greeks chewed parts of the mastic tree including the dried sap and bark, the Chinese chewed ginseng roots, the betel nut was chewed across South Asia, and both coca leaves and tobacco leaves were chewed long before people knew how to refine them into cocaine and cigarettes.

So it is hard to say when and where chewing gum truly originated. The ancient inhabitants of Scandinavia are the current claimants to the disputed title of the inventors of chewing gum – the oldest authenticated example is a 6,000-year-old piece of gum made from birch tar, which was found with tooth imprints in Kierikki in Finland.

# A Brief History of Sex and Sex Aids

When it comes to sexual behaviour and mores, history provides us with an astonishing variety of examples. While it seems that there is nothing new under the sun, and the examples below are simply the earliest that we know about, here are a few historical 'firsts':

- A 28,000-year-old phallus made of polished siltstone found in Germany is possibly the oldest known dildo found to date.

- Historical texts mention Egyptians and the early Greeks using unripe bananas, or camel dung coated in resin, for the same purpose.

- We know from references to the root of the mandrake plant – in Homer, the Bible and Egyptian texts – that aphrodisiacs date back at least 4,000 years.

- Phalluses have been found made out of stone, leather, wood and tar, although some of these may actually have been fertility symbols or good luck charms.

- An Egyptian papyrus from 4,000 years ago has the earliest known description of contraceptive methods. It involved a rather basic method in which the woman used a barrier of honey, gum or crocodile dung.

- While prostitution is known as 'the oldest profession in the world', the earliest evidence we have for it comes from Babylonian texts in the second millennium BC, when women had a duty to go to the temple and have sex with the first stranger to donate a silver coin.

- A Babylonian tablet from about 700 BC describes the use of a herbal pregnancy test.

- Commercial brothels date at least to 570 BC, when the *Porneion* of Athens opened for business.

- Brothels were also common in ancient Rome, but the first evidence we have of organized red-light districts comes from China in the Sung Dynasty (AD 960–1279), when groups of wine-houses that also served as brothels were run by the government's Imperial Board of Revenue.

# The Alphabet

## First Invented: Egypt
## Date: 2700 BC

The various alphabets in use in the modern world have complex histories. Their roots often come from purely pictorial (ideographic) systems of symbols, in which the picture directly represent objects. Over time some syllabic or phonetic symbols (in which the symbol stood for a sound) were added to the range of symbols, and it gradually became apparent that these were more flexible and useful as a basis for writing. An example of this evolution comes from Linear A and Linear B, two scripts used in the Mediterranean. Linear A was used by the Minoan civilization by about 2000 BC. It contained hundreds of signs, which were mainly ideographic, although some may have also represented phonetic sounds. Linear A is still undeciphered, but its descendant Linear B, which dates to five to ten centuries later, has been deciphered as a form of ancient Greek and contains just eighty-seven syllabic signs and one hundred ideographs. This is not as simple as a modern alphabet, but it is evidence of progress in that direction.

When it comes to the modern Western alphabet, the history is a tangled one. One possible chain of influence goes back to Egyptian hieroglyphs, which, by around 2650 BC, had been supplemented by twenty-two syllabic symbols used in combination with ideographs. One theory is that these symbols developed into an alphabetic system (known as the Proto-Sinaitic script) in the Middle East in about 1700 BC. This may have gradually changed into the Proto-Canaanite script, which in turn eventually mutated into the alphabetic Phoenician writing system. But this is all speculation, and the exact links are not known.

We do know for certain, however, that the Phoenician system was used widely around the Mediterranean area in about 1000 BC, that the Greek alphabet was a modification of it, and that the Romanized version of Greek became the basis of the Western alphabets used across most of Europe. So while the early part of the story is murky, we can trace the prehistory of our alphabet back at least to the Phoenician traders who spread it around the region, and possibly all the way back to Ancient Egypt.

# Playing Cards

## First Invented: China
## Date: AD 1000

It has long been known that playing cards were in use in China by AD 1000, about four centuries before they arrived in Europe. How they travelled from China to Europe has been a source of debate. While the Chinese cards were smaller and harder to shuffle, the resemblance is close enough that it seems undeniable that they influenced the design of the later European cards. Some theories that have been advanced include the idea that Marco Polo brought them back, that they reached Europe via the crusaders, or even that they were brought in by gypsies. The latter theory seems to have been based on the idea that tarot cards (used by gypsies for fortune-telling) may have pre-dated the game-playing cards. However, evidence suggests that the first tarot cards come from the middle of the fifteenth century in Italy, whereas the first playing cards had arrived considerably earlier. For instance, in 1377 the town council in Florence attempted to ban the playing of a card game called naibbe, which was something of a craze and was thought to be undermining public morals.

A discovery made in the archives of the Topkapi Museum

A woodcut Chinese
playing card, *c.* 1400 AD.

in Istanbul in 1938 seems to have settled the debate. The
collection holds a set of fifty-two cards, from about AD
1400, from the Mameluke Empire in Egypt. Similar to
the Western-style packs, there are court cards with *malik*
(king), *Na-ib Malik* (governor), *and Na'ib Thani* (deputy

governor). This probably explains the origins of the game *naibbe*. This set, and the subsequent discovery of cards from the Mamelukes a century earlier that are more obviously halfway between the ancient Chinese style and a more modern-shaped card, remove any doubt that playing cards came to Europe via Egypt.

The subsequent history of the suits is interesting. The Mameluke cards were divided into depictions of coins, swords, goblets and polo-sticks. Polo was unknown in Europe, so cardmakers replaced the sticks with batons and the goblets with simpler cups. Italian and Spanish cards still have sticks, swords, cups and coins as the suits today. In the fifteenth century, German cardmakers adapted this to a more naturalistic scheme – one of acorns, leaves, hearts and bells. The French modified this in about 1480 into simpler shapes that could be printed with stencils: *trèfle* (clover), *pique* (pike-heads), *carreau* (paving tiles) and *coeur* (hearts). The final English version derives from multiple sources – spades are based on the earlier swords and pikes; clubs are based on the Spanish suit of staves (or sticks), diamonds come from the shape of the paving tile (but also reflect the wealth of coins); and hearts is retained from the French version.

# The Garden

The earliest gardens developed sometime between the birth of agriculture (about 10,000 years ago), and the *Epic of Gilgamesh* (about 4,000 years ago), which describes a sacred place: 'With crystal branches in the golden sands, In this immortal garden stands the Tree, With trunk of gold and beautiful to see.' As this quote suggests, early gardens were often inspired by the widespread veneration of trees as religious symbols. The Garden of Eden is a similar ancient piece of mythology, in which a safe space where trees grow is superior to the barren ground beyond.

Developing from enclosures within a farm, in which fruit trees created a pleasant area of shade, they also came to be seen as places of recreation and rest. In the early gardens of the Mediterranean and Near East, there was still more focus on trees than on flowers. The word 'paradise' is etymologically connected to the ancient Persian word for garden – a paradise was a lush area with water running in channels or fountains, and a regular pattern of trees. This style spread from its origins in the fertile crescent of Persia, Egypt and Assyria, and was adopted across North Africa (from where it spread to Europe) and even in India, where the Mughal gardens show the same influence. As a result the enclosed garden, often with paths dividing it into quarters, became both a place of seclusion and leisure and a symbol of religious imagery across a large part of the world even before Christianity and Islam co-opted this idea into new religions.

A scene from sixteenth-century Mughal India, where garden design reached a high point.

# Glue

### First Used: Central Italy
### Date: 200,000 years ago

The earliest evidence of adhesives being used by early humans or Neanderthals comes from Italy, where a 200,000-year-old pair of stone flakes were found with the remnants of birch-bark tar. This is characteristic of stone tools that were hafted (glued to a wooden handle).

The first indication we have of compound adhesives being used in prehistory comes from Sibudu, South Africa, from 70,000 years ago – again this comes from stone segments of axes that had been hafted to handles. The adhesive remnants are made from plant gum mixed with red ochre (iron oxide), which makes the gum stronger while protecting it from a wet environment.

By the fifth millennium BC some more advanced processes were being used in Europe and the Middle East. Animal glue, for instance, was first used in ancient Egypt. Manufactured by prolonged boiling of animal hides and hooves, types of glue were created that could be used to reinforce papyrus scrolls, and to create expensive furniture (examples of which have been recovered from the tombs of pharaohs, including that of Tutankhamun).

# Money

## First Used: Mesopotamia
## Date: 3000 BC

The prehistory of money is fascinating from a philosophical point of view. It can be presumed that from the early period of agriculture at the latest, some crops and goods were traded or bartered. For instance, a cow might have been swapped for a box of seeds or for a day's labour on the land. It seems obvious that some goods would fairly early have become useful as stores of value – for instance, in the barter of seeds and the cow, the seeds might be useful to keep for future exchanges, since they are relatively durable and can be subdivided. Since pure barter only works where two traders have exactly the goods the other party wants, goods such as the seeds would very quickly have become a kind of proto-money (generally called 'commodity money').

Using tally sticks or other primitive records of what had been bought or sold, people could also grant or store credits for later purchases. This means that the concept of debt was also part of monetary systems from an early stage. In fact, the concept of negative numbers was invented by Chinese mathematicians who wanted to record both credits and debits. They recorded debits in a different colour to credits, and subtracted them rather than adding them.

Relative rarity also made an object suitable for use as commodity money. For instance, cowrie shells were in use as money tokens 3,000 years ago on the Indian subcontinent, because it was hard to acquire an excessive quantity of them.

It is uncertain when a good that was being used in such exchanges first became seen specifically as 'money', but the earliest organized currency we have records of comes from the Mesopotamian civilization in about 3000 BC. The unit of weight and currency was the shekel, which referred to a particular weight of barley, and equivalent amounts of metals such as silver, copper and gold. The fact that the shekel was defined in this way suggests that both barley and precious metals had already been in use as commodity money, and that this was an attempt to legislate how they were used and to define their relative value.

The Code of Hammurabi, the best-preserved ancient body of laws from Babylon in about 1760 BC, gives us a sense of the role that money played in these civilizations. It lists fines and compensations for malpractice, and sets a limit on how much interest can be charged on a debt. Even the earliest of civilizations had problems with sharp practice among moneylenders!

# Ancient Matches

## First Invented: China
## Date: AD 577

Throughout history, mankind has looked for ways to control fire for the purpose of heating and cooking, as well as for a variety of military and industrial purposes. Early man learned to rub sticks together or to strike sparks from a stone, and also learned techniques for keeping hay or straw smouldering for long periods and transporting it from place to place. But all of these ways of starting a fire were cumbersome, and it was obvious that an easier technique would be a big step forward.

The first true self-striking match was invented fairly recently. Following on from experiments with phosphorus by the seventeenth-century alchemist Hennig Brandt, the French inventor Jean Chancel created a rather dangerous-sounding match in 1805. A stick was coated in potassium chlorate, sulphur, sugar and rubber, and then dipped into a glass bottle of sulphuric acid. The ensuing chemical reaction made the wood ignite, giving off noxious fumes as it burned.

However a simpler form of match had been in existence in China as early as the sixth century. The story is that the court women of the Northern Qi (a kingdom that was only briefly in existence) invented the match during a siege in which there was a timber shortage and it was necessary to preserve wood for cooking and heating. A short pine stick was dipped into sulphur. Two of these sticks had to be rubbed together in order to ignite one or both of them. This invention spread rapidly around China, where it was used for lighting stoves, lamps and fire crackers, among other things. The poet T'ao Ku lyrically described the matches as 'light-bringing slaves' in a tenth-century text, and they were also sold as 'fire inch-sticks'.

It is thought that these matches may have been brought to Europe by Marco Polo or a contemporary traveller to China, as we know that they were sold in markets in Hangzhou, which he visited in the thirteenth century, but there is no confirmed record of such matches in Europe until AD 1530.

# Rubber

## First Discovered: Mesoamerica
## Date: 1600 BC

Rubber (or caoutchouc) is a natural product that can be tapped (drained) from the rubber tree, a native of South America. It was discovered and used by the indigenous cultures of the area as long ago as 1600 BC, in the Olmec civilization. One use was to make balls to be used in their ritual ballgames. The Mayans and Aztecs also used rubber for purposes such as waterproofing textiles and making containers. The Olmec culture also knew how to stabilize rubber, through some unknown combination of heating and added ingredients.

Rubber was introduced to Europe in 1736 via an academic paper by Charles Marie de la Condamine in France. In 1770, the English scientist Joseph Priestley discovered that it could erase a pencil mark and this inspired its modern name of 'rubber' (something that rubs). However, rubber was scarce throughout the early nineteenth century, as it was a capital offence to export seeds from Brazil, the main source at the time. Eventually the explorer and bio-pirate, Henry Wickham, managed to smuggle a large number of seeds back to England, where a proportion of them grew into trees under the care of Kew Gardens. Samples were sent all over the British Empire and rubber plantations were created in many countries including India, Singapore and British Malaya. Rubber came to be used in building, cars, flooring, car tyres, gloves and balloons, and much more.

As a sidenote, Charles Goodyear is generally regarded as the inventor of vulcanization, the process by which sulphur or other substances are added to rubber as it is heated in order to create a much harder type of rubber. It is not so often remembered that the Olmecs had beaten him to their version of the same process over 3,500 years earlier.

## Mirrors

Still water has been used as a mirror since the earliest parts of history, as we know from the myth of Narcissus falling in love with his own reflection in a pool. To make a mirror you need a flat, extremely smooth surface, with high reflectivity. The earliest mirrors that have been found are made of obsidian, a black volcanic glass, and include examples from about 6000 BC in modern-day Turkey. Polished copper mirrors have been found from about 4000 BC in Mesopotamia and about 3000 BC in Egypt. In Central and South America, polished stone mirrors were developed before polished metal ones (stone versions have been found dating to 2000 BC). Some early Chinese mirrors from the same period were made of polished bronze and copper by the Qijia culture, who lived in the upper part of the Yellow River region.

# Silk

## First Invented: China
## Date: 3000 BC

Sericulture, the cultivation of silkworms to make silk, is an example of extraordinary ingenuity. The secret of how to make it was jealously protected and confined to China for millennia – as late as the first century BC, the Roman historian Pliny believed that it was made by 'removing the down from leaves with the help of water'.

Its discovery goes so far back in time that the people involved are legendary. Lady Hsi-Ling-Shih, 'the wife of the Yellow Emperor', who supposedly ruled in about 3000 BC in China, is known as the 'Goddess of Silk' as it is claimed that she initiated the first silk production. We do have evidence that sericulture was practised 5,000 years ago. Half a silkworm cocoon showing signs of the production process, dating back to about 2600 BC, has been found in the soil beside the China River, while a group of ribbons, threads and other silk fragments found in Zhejiang province dates to about 3000 BC. More recent finds have suggested that the secret was known even earlier – a small ivory cup decorated with a silkworm motif seems to be about 6,000 to 7,000 years old, while other remnants of sericulture from the lower Yangtze River suggest it might even have been happening before 5000 BC.

The actual secret of sericulture is the lifecycle of the blind, non-flying moth *Bombyx mori*. (Historically, silk production may have started out using the wild moth *Bombyx mandarina*, which probably evolved into the current species, and there are also a few other related species now used.) The moth lays a batch of 500 or so eggs shortly before it dies. From these tiny pinpoint-sized eggs come about 30,000 worms, which eat huge quantities of mulberry leaves as they grow to 10,000 times their original size, and produce twelve pounds of raw silk as they create their cocoons. Each cocoon consists of a single thread of up to 900 metres (984 yards) in length. When the time comes for the cocoons to be farmed, they are subjected to heat (either using steam or a baking process) to kill the pupa. The silk is wettened to loosen the thread, which is carefully unwound and dried.

The Goddess of Silk is also supposed to have been the inventor of the loom, which is mentioned in ancient Chinese texts. However, the oldest looms that have been found intact are only about 2,200 years old, so how early in history looms were actually used and when the pattern loom was invented remains unknown.

The famous Silk Road became known as a trade route along which ancient travellers transported valuable goods such as jade, horses and ivory. But of course the most significant item was that for which it was named, the remarkable material from China which came to be coveted around the world for its softness and elegance.

# Early Games

## First Invented: Prehistory

Games and play are an integral part of human experience going back into our distant evolutionary ancestors – the children of many mammals engage in forms of play as part of their learning experience. So the challenge for the historian is to establish when particular games originated. Even here the evidence can be frustratingly difficult to identify. Most archaeological digs come up with a variety of small objects such as round balls of clay, shaped bits of wood and so on, but it can be hard to establish whether these are part of a formal game, ritual objects, or something else altogether.

The Greek historian Herodotus, writing in the fifth century BC, credited the ancient Lydians with the invention of dice, knucklebones, ball games, and 'all other games of this sort except checkers'. The claim is that there was a serious famine in the thirteenth century BC and the Lydians came up with the games to distract themselves and pass the time. It's a nice story, but unfortunately it is clearly wrong. Dice, for instance, have been found in the Indian subcontinent that were being used at least 1,000 years earlier than the Lydian famine. Both six-sided dice and four-sided dice have been found from this period, often made of bones and knucklebones in particular.

The oldest known artefacts that seem to be part of a game are forty-nine carved painted stones from the Başur Höyük burial mound in Turkey (which is about 5,000 years old). From a similar period in ancient Egypt, we have *Mehen* boards, shaped like a coiled snake divided into rectangular spaces – Mehen was a mythological snake deity. The rules of the game are unknown to us. Other objects have been found in areas of Syria and Iraq, and from later cultures such as the Akkadian and Babylonian civilizations.

One early game for which the rules are known is the Royal Game of Ur, a racing game with a set of pieces on a marked board played with dice. The game survived and was played by ancient Egyptians many centuries later. The pure dice game that the Romans called *Lusus Duodecim* had also probably been passed down from antiquity.

An early version of draughts or checkers was also played by the ancient Egyptians. The game of *senet* was played on a board with thirty squares (arranged three by ten). The goal was to reach the far end of the board. Other toys found in ancient Egyptian sites have included a toy crocodile that snaps its mouth, and wooden dolls that appear to be diversions rather than ritual objects, for instance a baker who moves to knead his dough made out of wood.

By the time of the Greeks and the Romans, many early

The game of Senet being played, c. 1298–1235 BC.

versions of modern games had been developed, including handball, knucklebones, tic-tac-toe and a variety of board games. While some of these board games had different pieces which took on different roles in, for instance, a war reconstruction, the game of chess came somewhat later. It is known that it existed in India by the sixth century AD, and most historians suggest that this was where it was invented, although it has also been suggested that it came to India from China.

Dominoes, which is related to dice in its use of the numbers one to six on the pieces, is commonly supposed to have originated in Italy, although the earliest historical reference to it appears to be from the Chinese state archives which record a set of 'thirty-two pieces marked with 227 pips' as being having given to the emperor as a gift.

## The First Skiers

The earliest skis, which would have been used to enable journeys across snowy and icy landscapes rather than for sport, were made of wood. As a result, very few remain. However, the acids in a peat bog at Vis in northeastern Russia (near the Ural Mountains) preserved a few fragments of a wooden ski. The most impressive section is the front end, which is decorated with a carved elk's head. This might also have functioned as a brake.

The earliest image we have of a person on skis comes from a tiny island in the north of Norway. A rock carving from about 4,500 years ago depicts a figure on skis that appear to be about twice his length. Using a single stick to steer, and crouching in the now-standard downhill ski-ing posture, he appears to be wearing some kind of costume, as he has ears like a hare. This might be a hunter's good luck charm.

## CHAPTER 2

# MECHANICAL AND INDUSTRIAL TECHNOLOGY

# The Extraordinary Steam Engine of Hero of Alexandria

## First Invented: Greece
## Date: First century AD

Hero of Alexandria (AD 10–85) was a brilliant teacher of mathematics and physics at the Musaeum (which encompassed the famous Library of Alexandria). He shared Archimedes' knack of converting his academic genius into the creation of remarkable devices. His inventions included the world's first vending machine, which provided a measure of holy water in return for a bronze five-drachma coin; a wind-powered organ; the force pump; and a famous stand-alone fountain that was named after him.

He also created a set of self-opening doors that were used by temple priests to create a sense of wonder. When a fire was lit, the water in a concealed metal globe heated up. It was then forced through a tube into a bucket, whose weight pulled the doors open via a system of pulleys. When the fire was extinguished, the water in the system cooled and was sucked back into the globe, whereupon the bucket rose and the doors closed themselves.

A number of these inventions relied on the transference

of weight or the use of water within an enclosed system. The vending machine worked because the weight of the coin pulled a small pan downwards and opened a valve, which allowed the holy water to escape. Hero's stand-alone fountain was powered by hydrostatic energy (the buoyancy factor that makes objects feel lighter in water), and the temple

The *aeliopile* is considered to be the first recorded steam engine.

doors worked because of the expansion and evaporation of heated water. Hero's most famous invention – the aeolipile – used some of these elements in a novel way to create a device that can reasonably be described as the world's first steam engine (and which also prefigured the jet engine).

The aeolipile consisted of a closed cauldron containing water, which was heated by a fire. As the water heated up, the steam rose through two pipes that acted as the axle for a globe mounted between them. The globe had two outlets, through which the steam could escape. The escaping steam forced the globe to rotate rapidly on its axle. (One recent recreation of the aeolipile rotated at an astonishing rate of 1,500 rpm, which is impressive even when compared with modern steam turbines.)

Hero's engine, as it is commonly known, was not a practical device for generating energy as it was simply too inefficient – it was created more as a mechanical marvel than for the purpose of industry. However, the same technology could in theory have been combined with cylinders and pistons (which Hero utilized in other devices, such as his firefighting water pump) to create a more viable energy source. It is unclear why Hero didn't take this next step. Some have speculated that, given the widespread slavery and the ongoing threat of warfare in this period, inventors tended to look for impressive illusions, recreational devices or military applications, rather than seeking to replace the ready supply of slave labour.

To understand how remarkable Hero's achievement was, we need to bear in mind that true steam power wasn't developed until the seventeenth century. Jerónimo de Ayanz y Beaumont patented a basic steam-powered water pump in 1606; Thomas Savery's 1698 steam pump used condensing steam to pressurize steam to pump water; the atmospheric engine of Thomas Newcomen, which came into use in 1712, powered pumps for the mining industry; and it wasn't until 1781 that James Watt invented the first steam engine which could power continuous motion. So Hero richly deserved the accolade of his nickname *Michanikos*, the Greek word for 'Engineer'.

# The Pharos Of Alexandria, The World's First Known Lighthouse

**Built: Alexandria, Egypt**
**Date: Third century BC**

The lighthouse was probably a Greek invention, although the Romans adopted it with enthusiasm and surrounded their empire with a network of impressive structures. Homeric legend credits Palamadis, of the city state of Nafplio (in the Peloponnese), with the original invention of the lighthouse. (He is also reputed to have invented weights and measures.) And we know that the Athenian politician Themistocles was responsible for the construction of a fire beacon on a raised stone column at the entrance to Piraeus, the port in Athens, in the fifth century BC. Elevating the fire meant that it lit the immediate area more efficiently and could be seen from further out at sea.

The best-known ancient lighthouse is the Pharos of Alexandria, which is remembered as one of the Seven Wonders of the World. Alexander the Great founded the city on an isthmus opposite the small island of Pharos. The gap between the city and island was filled with a mole – a huge breakwater made of stones. The mole was known as the *Heptastadion,* meaning 'seven stadia'. A stadium was a unit of length of approximately 180 metres (600 feet). This created a massive enclosed harbour for the city.

In the period after the death of Alexander, a giant lighthouse was constructed on the island. Work commenced in 280 BC and the structure was complete in 247 BC. The limestone tower was 106 metres (348 feet) high, and consisted of a massive square base on which a second octagonal section supported a circular final section at the top. This last tower contained a burning furnace that could reportedly be seen from as many as 160 kilometres (100 miles) away. During the day, a giant mirror was positioned to reflect the sun's rays out to sea for the same purpose.

The purpose of the lighthouse was to guide shipping safely into the harbour. However, its magnificence also made it a gigantic status symbol, which would become famous around the world. The name of Pharos soon came to be used for the lighthouse itself rather than the island on which it stood. Several modern languages have words for lighthouses that derive from it: *pharos* in Greek, *faro* in Italian, *phare* in French, *far* in Romanian, and in Russian, the related word for headlight, is *fara/фара*.

The lighthouse survived intact for over a millennium. It was damaged by earthquakes in AD 956, 1303 and 1323, and

The Lighthouse of Alexandria, Egypt, is estimated to have been 100 metres tall (constructed in 280 BC, destroyed in 1480 AD).

the last remnants were converted into a fort on the same site in 1480. There is also an apocryphal story that the initial damage was caused in the tenth century when a Byzantine spy won the trust of the Egyptians and was given permission to dig for secret treasure on the island. His excavations are reputed to have been so cunningly carried out that the foundations of the massive lighthouse were fatally undermined.

# A Very Brief History of Metalwork

The ability to fashion metal into tools and ornaments played a fundamental role in the development of human civilization. Here are a few of the breakthroughs that were made by ancient cultures.

- The first metals to be used by man were copper, gold and meteoric iron (iron that could be extracted from meteorites).

- The first tools were made of copper, which was being used by about 9000 BC.

- A copper pendant found in Iraq has been dated to 8700 BC.

- Stone Age man was as fascinated by gold and silver as we are today, and had learned to shape these metals into jewellery by 6600 BC.

- Copper was initially used in its native form. However, copper smelting was discovered independently in

China and Europe in approximately 3500–3000 BC.

- Ötzi the Iceman (found in the European Alps), who died in about 3200 BC, had traces of arsenic in his hair and a high-grade copper axe, which suggests that he worked in copper smelting.

- Tin mining developed from 3000 BC onwards. The oldest tin-mining area that has been identified is Erzgebirge, on the border of Germany and the Czech Republic.

- The Bronze Age began in about 2300 BC, when people discovered the art of metallurgy (mixing two metals together to create a stronger one). Bronze is made by mixing copper and tin.

- By the third millennium BC, palaces in the Mesopotamian civilization displayed hundreds of kilograms of finely worked copper and gold, and they were also using bronze for weapons.

- Gold was being worked with precision in Peru in South America by 2000 BC, and in 1400–1000 BC craftsmen in the Olmec civilization were able to polish iron to the point where it could be used as a mirror.

- The ancient Egyptians of the same period used the old technique of fire-setting (heating rock with flames in order to be able to shatter it), as well as mines, to extract large quantities of gold.

- The art of metallurgy did not reach South America until the Middle Ages, but once it did, the Aztec, Inca and Mayan civilizations developed highly sophisticated ways to use alloys and gilting techniques.

- The ancient Greeks were the first to use bronze casting.

- While cast iron and wrought iron had been in use for millennia, the Iron Age, which started in about 1200–800 BC, was characterized by large-scale production of steel, a reinforced form of iron.

- A technique for making steel was discovered independently in the Tanzania region 2,000 years ago. The Hayan people used open-air furnaces of mud and grass to provide the carbon that is needed to transform iron into steel.

# Greek Technology

The period from 300 BC to AD 150 in the ancient Greek civilization is notable for having produced many devices that would remain in use throughout the Western and Muslim world for over 1,000 years. Examples include screws, organs, dials, the speedometer, diving equipment, parchment, roof tiles, breakwaters and many more.

The four means of non-human propulsion that were discovered by the ancients were: watermills, windmills, steam engines and animal power. It was the Greeks of this period who pioneered the first three of these. As well as Hero and Archimedes, notable inventors included Ctesibius (285–222 BC), known as the 'father of pneumatics', who used pneumatics to invent a powerful water pump (see page 32) and to create the world's first pipe organ. The design of modern piano keyboards would eventually be based on his original organ.

# The Spoked Wheel

## First Invented: Russia
## Date: 2000 BC

We think of the wheel as one of the most significant inventions of antiquity, but its development was fairly slow and gradual. Stone Age people had already realized that heavy objects could be moved more easily if they were placed on rolling logs, and this may have been the spark that eventually led to the wheel. The potter's wheel was in use about six millennia ago. We have, for instance, examples from the Mesopotamian area from about 3500 BC.

The first clear evidence we have for wheeled vehicles comes from the same period. A pot dated to 3500–3350 BC, from the Funnelbeaker culture of southern Poland, depicts a wagon with four wheels and two axles. Stone wheels were used at first, but it became apparent fairly soon that a lighter wooden wheel was preferable.

However, even a wooden wheel is quite a heavy object, and the early vehicles were difficult to move over uneven or wet ground. So the real breakthrough in wheel construction came with the spoked wheel. The earliest known example comes from the Sintashta culture of the Russian steppes in about 2000 BC, while the horse-riding tribes of the

Caucasus were building war chariots with spoked wheels over subsequent centuries.

Wheelwrights of this period started from a solid piece of wood and used a special tool called a spokeshave to cut out roughly triangular sections of the circle, leaving an outer circle, an inner circle (with a hole for the axle) and spokes that joined the two circles and reinforced the structure. The result was a much lighter wheel, which was nonetheless durable. This method spread gradually through Europe, with the modification in some areas (including the Celtic cultures) of an iron rim.

A chariot wheel found in an Egyptian tomb, c. 1400 BC.

Of course, once wheels existed, it created a more urgent need for even, well-built roads. It would be the Romans who took that art to new highs, from about 500 BC onwards. But on the relatively even ground of the steppes and the Caucasus, the spoked wheel was being put to effective and deadly use long before the Roman road became famous for its virtues.

# Prehistoric Inventions

The wheel was an important development in human history, but it was far from being mankind's earliest breakthrough. Several important inventions had been in use for thousands of years before the wheel.

The earliest evidence for **sewing needles** comes from over 60,000 years ago – a fragment that appears to be the sharp end of a bone needle found in Sibudu Cave, South Africa. A bone needle from about 45,000 years ago was found in Potok Cave in Slovenia. Bone and ivory needles have also been found in China and Russia that date back to about 30,000 years ago.

**Clothes** may have been worn for as long as 170,000 years. The current evidence for this is based on the scientific hypothesis that the human body louse, which lives in clothing, diverged in its evolution from the head louse at around that time, suggesting that humans were wearing some form of clothing. As well as the evidence from sewing needles mentioned above, the oldest known dyed flax fibres (made from linseed) come from a cave in Georgia and date back to 36,000 years ago.

**Rope** of one sort or another has been used since prehistoric times. The very first ropes may well have been sections of plants like vines, and it is likely that braiding and twisting to strengthen these were the first type of ropemaking. Fired clay with the impressions of rope fibres has been dated to 28,000 years ago, and fossilized fragments of two-ply rope from about 15,000 BC were found in one of the Lascaux caves in the Dordogne.

The earliest **baskets** that can be accurately identified with carbon dating come from about 10,000 to 12,000 years, although it is likely that the art was known somewhat earlier. As they are generally constructed from plant or animal material that decays, very few have survived.

The oldest surviving **boat** is the Pesse canoe. It was made from the hollowed-out trunk of a pine tree in about 8000 BC, and was found during roadbuilding in the Netherlands in 1955. However, the settlement of Australia about 40,000 years ago and other pieces of archaeological evidence suggest that boats or rafts were being used to cross significant bodies of water far earlier in human history.

Finally, **music** is one of the most ancient human activities. **Flutes** are the oldest instruments that we know of. Examples with hand-bored holes have been found in Geisenklösterle cave in Germany that date back to about 40,000 years ago. The earliest flutes were made of bone or ivory, such as mammoth tusks and swan or bear bones.

# The Hot Air Balloon

The Chinese were using unmanned hot air balloons from about the third century BC. One ancient story claims that the military strategist Zhuge Liang (also known as Kongming) sent a sky lantern into the sky to call for help when he was surrounded in battle. As a result (or possibly because the lanterns resemble the hat in which he is often depicted) they came to be known as Kongming lanterns. It would be perfectly possible to build a manned hot air balloon using ancient materials, but there is not a great deal of evidence to sustain the theory that this happened. We do, however, have at least one mention in an old document that the Chinese had 'solved the problem of aerial navigation', which might suggest that they did also send manned balloons into the air.

It is also worth mentioning the artworks depicting creatures on the Nazca plain, which were created by the Nazca culture between AD 500 and 900. The size of these artworks means that they are difficult to view in their entirety from any point on the ground, but they can be observed to spectacular effect from the air. Some have argued that this proves the Nazca had a means of flying, and had thus independently invented the hot air balloon. (Although we should also note that the artworks have also been advanced as proof that earth was colonized by 'ancient astronauts' from another planet, so we should be careful about leaping to any wild conclusions.)

Aerial view of geoglyphs near Nazca, Peru.

# The Crane

## First Invented: Greece
## Date: Sixth century BC

The wheel is one of the traditional six 'simple machines' that were listed in antiquity by writers such as Hero of Alexandria. These were the basic devices that allowed people to exert more force than they could manage on their own. The other simple machines are the lever, the wedge, the inclined plane or ramp, the pulley and the screw (see page 82). Of these, the lever and the wedge were used from the Stone Age onwards, while the ramp was certainly in use by the construction of the Great Pyramid in Egypt (2600 BC), and probably much earlier than that for simpler projects.

Moving heavy weights with a ramp is, however, an inefficient method that requires concentrated amounts of manpower, and the invention of the pulley was a major step forward. It is thought that rope pulleys were being used in Mesopotamia in about 1500 BC, but the first definite evidence of their use comes from ancient Greece. Archimedes (c. 287–212 BC) wrote about how compound pulleys were used in a block-and-tackle system. This increases the efficiency of the pulley, which means that less force is needed to lift the same weight. Stone blocks used in

Greek temples from the sixth century BC have marks and holes that indicate that a crane must have been part of the construction process.

The simplest Greek crane was the *trispastos* (three-pulley crane). This consisted of a jib (a tilted strut made up of two joined masts, over which the rope was slung), a winch (a lever which turned and pulled the rope), and a block with three pulleys. This gave the operator a three-to-one advantage, meaning that a single worker could lift up to 150kg. Cranes were used in the construction of impressive buildings such as the Parthenon. For heavier loads, the Greeks might use a version with five pulleys (*pentaspastos*) or even a more complex set of three-by-five pulleys (*polyspastos*).

The Romans adopted the crane and used them to lift even more extraordinary weights. Some of the blocks at the temple of Jupiter at Baalbek, which weigh as much as 100 tons, had to be lifted to a height of about 19 metres (62 feet), which would have required a crane considerably taller than that.

Roman *trispastos* crane, which enabled a single man to raise very heavy loads.

# The Screw

The last 'simple machine' to be invented was the screw, reputedly by Archytas of Tarentum (428–350 BC). Screws were originally used to exert pressure on olives and grapes in order to extract the oil and juice. A significant further development came with the invention of the Archimedes screw, which was either invented or popularized by Archimedes. (There is evidence of its use in Egypt, and possibly in the water system of the Hanging Gardens of Babylon before this time.) This relied on the same principle as the screw, but used the force generated as a way of raising water. The simple rotation of the screw drives the water upwards with fairly minimal exertion. It was initially used to irrigate land and to drain bilge-water in ships and boats. The Romans extended its use to the drainage of mines, and for centuries afterwards it remained the most efficient way of continuously lifting water.

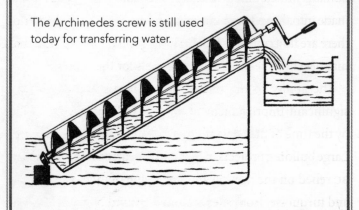

The Archimedes screw is still used today for transferring water.

# Tunnels and Mines

## First Used: Swaziland
## Date: 43,000 years ago

The search under the surface of the planet for particular types of stone and metal started in prehistory. Flint, which was used for weapons and tools, was one of the first stones to be mined. Seams of the stone on the surface would gradually be mined and hollowed out, and Paleolithic man soon learned to follow the seams further into the surrounding rocks. By the Neolithic period (from about 12,000 years ago onwards), there were significant numbers of mines and tunnels around the world. The oldest we have found is in the Lion Cave in Swaziland, which has been dated to about 43,000 years ago. Paleolithic humans mined a mineral called hematite here, which can be made into the red pigment called ochre. From a similar period there are flint mines in Hungary, suggesting that Neanderthals may also have been able to dig tunnels for their tools.

The ability to dig tunnels down into the rock led to a significant improvement in humankind's tunnelling ability by the time of the ancient Egyptian and Babylonian cultures. Large building projects in Egypt in the early third millennium BC relied on the mining and quarrying of malachite, copper and turquoise, from sites scattered around North Africa and the Middle East. The gold mines of Nubia were amongst the most extensive of ancient times. By 2180–2160 BC, the Babylonians had the technology to dig a tunnel of 3,000 feet (914 metres) in length that passed under the Euphrates river.

One of the most fascinating archaeological finds of recent times is the existence of hundreds of tunnels in the rock under Neolithic settlements in most parts of Europe. Hundreds of these have survived to the present day: archaeologists have recently estimated that there must have been thousands of them across Europe, from northern Scotland to the shores of the Mediterranean. These tunnels are large enough for a person to crawl through, with occasional chambers and rooms. It is unclear whether they were used for travelling safely from place to place or as a refuge, but the effort that went into digging them out of the rock suggests they must have been highly important.

# Oil Wells and Boreholes

**First Invented: China**
**Date: Third century AD**

Drilling holes and extracting liquids from them is a more complex task than mining for rocks and minerals. By the

Han Dynasty (AD 202–20), Chinese mining engineers had come up with a clever technique for drilling holes that reached as far down as 600 metres (2,000 feet). A beam, or series of beams, pushed a drilling bit deep into the ground, while force was applied by workmen jumping onto a platform that forced the beam down. Meanwhile a team of buffalo or oxen pulled an arm that rotated the boring tool. (It's interesting to note that the first wave of petroleum extraction in California in 1860 used an almost identical method, before more advanced technology could be brought to the region.)

By AD 347, there were also oil wells being dug in China, using drill bits attached to bamboo beams. The oil was burned for a variety of industrial purposes, such as the conversion of brine into salt. There are records from the tenth century of networks of bamboo pipelines connecting salt flats with the oil wells. Both China and Japan were using a form of petroleum and natural gas from the seventh century.

Elsewhere, the Persian alchemist Muhammad ibn Zakarīya Rāzi seems to have been the first to apply the art of distillation (see page 28) to petroleum, producing kerosene in the process. From this point onwards Arab and Persian craftsmen worked out how to distil crude oil in the production of other flammable substances that could be used in military projectiles.

# Glassmaking

### First Invented: The Fertile Crescent
### Date: Third millennium BC

The fertile crescent stretches from the northern end of the Nile river up through Syria, then back down through Iraq to the Persian Gulf. It is known as one of the earliest cradles of civilization, as it was here that the Sumerian, Mesopotamian, Babylonian, Akkadian and ancient Egyptian cultures developed. The earliest pieces of manufactured glass that have been found come from various parts of the fertile crescent – although it is believed that the Egyptians may have been the first to develop the craft from some time in the third millennium BC onwards.

Glass, in the form of obsidian, occurs naturally in some places, and was a highly prized substance that was traded around the world in prehistory. However, it was the increase in metalworking that led to the discovery of glass production. The first pieces were probably small droplets that were a by-product of metalwork.

Glassmaking technology developed at a significant rate throughout the first half of the second millennium BC in the Egyptian area. Objects from this period include ingots, vessels and beaded necklaces. By 1500 BC, fine

A Roman 'cage cup', c. fourth century AD. These special beakers were status symbols – there are only fifty surviving examples. (See also the Lycurgus Cup on page 99.)

glass objects were relatively common, and the Phoenicians of the modern-day Lebanon had become well known for their glassmaking expertise. The first coloured glass (made by adding oxides to the molten glass) also dates from this period. Early vessels were 'core-formed' – a process in which a rope of heated glass was wrapped around a core of clay and then fused by being repeatedly reheated. The result was less smooth than blown glass, but nonetheless impressive. Coloured glass threads could also be wound around the outside of the vessel as extra decoration.

In the ninth century BC there was a resurgence of

glassmaking technology (after a hiatus that followed the collapse of several late Bronze Age civilizations). This was when colourless glass was first made.

One reason why our early knowledge of glassmaking is somewhat incomplete is that the craftspeople tended to be extremely secretive. Secrets of the craft were passed down the generations within families or guild-like organizations, but not shared widely, and any written records may well have been in codes or ciphers. The first known glassmaking manual was kept on cuneiform tablets found in an Assyrian palace from about 650 BC.

The next big leap in glassmaking came with the discovery of how to blow glass. The first known example comes from Jerusalem in AD 50. As well as allowing glassblowers to make much finer, thinner glass vessels, it also led to the creation of the first glass windows. The glass was blown into a sphere, which was then formed into a sausage-shape with the ends opened up to create a cylinder. This was cut down one side with a heated tool, and the resulting sheet was drawn out with pincers on a flat surface.

Windows became a status symbol in Roman homes – all the more so in the northern part of their empire, where the colder climate meant that windows were a big step forwards in the villas of the wealthiest citizens. But glass windows remained a luxury for many more centuries to come, only becoming standard parts of the normal house in the past century or so.

## Movable Type

The German Johannes Gutenberg tends to be credited with inventing the printing press, but the Chinese again beat the West. Movable type was first used in China in the ninth century. However, movable type didn't catch on in China due to a quirk of language – recombining the 80,000 characters of Chinese was simply not as efficient or timesaving a process as it was for European languages. Even after the introduction of Western printing presses from the sixteenth century onwards, woodblock printing remained more popular and convenient. It was not until the nineteenth century that movable type was more widely used in China and Korea.

# The Four Great Inventions

At one point it was widely believed in Europe that four of mankind's most significant inventions – gunpowder, the nautical compass, printing and papermaking – were first used in the West. It was only from the 1530s onwards that reports from Spanish and Portuguese sailors started to circulate, revealing that all of these inventions had in fact already been discovered in China.

The lodestone compass was first used during the Han Dynasty (206 BC–AD 220). Initially it was used only as a strange contraption for fortune telling. It's uncertain when its navigational potential was first realized, but there is a reference in a book (dated to about AD 1040) to an iron 'south-pointing fish' which can be used for finding the way. Later in the same century, the author Shen Kuo described a magnetic needle compass. However, the most common compass during this period was simply a magnetized needle floating in a bowl of water (a 'wet compass'), rather than a pivoting needle in a case (the 'dry compass').

It is possible that the South American Olmec civilization knew how to use wet compasses in about 1000 BC, but the evidence, a grooved lodestone that might have been floated in water to form a compass, isn't definitive.

Gunpowder was discovered by Chinese alchemists, looking for elixirs with magical properties, in about AD 850. Saltpetre, the common name for potassium nitrate (an oxidizing agent), was mixed with sulphur and charcoal, and the result was a strange substance that was prone to bursting into flames. After experimenting to find the correct, stable mixture, which would explode only when you wanted it to (and having presumably lost a few eyebrows and worse along the way), gunpowder was the result. It was initially used for 'flying fire' – a projectile in which a container of gunpowder was attached to the shaft of an arrow with incendiary results. This weapon, along with early types of grenades and cannons, was used by the

Chinese to great effect in their battles with the Mongols. Gunpowder spread gradually via the Islamic world to Europe, where it was first inserted into a handheld gun by Italian craftsmen in the late fourteenth century.

It used to be thought that papermaking was invented by Cai Lun, an official of the Imperial court in China, in about AD 105. He made a sheet of paper from various plant fibres, including mulberry, fishnets and pulped hemp. However, a more recent archaeological discovery suggests that paper was already being made about one hundred years earlier. It was initially used for wrapping and padding. Other inventions that subsequently derived from it in China include writing paper (from about the third century), toilet paper (from the sixth century), teabags (from about the seventh century), and paper currency (first issued in the Song Dynasty, which lasted from AD 960 to 1279).

Finally, printing had been discovered by the later stages of the Tang Dynasty (AD 618–907). The first dated book that has been identified is a copy of the *Diamond Sutra*, from AD 868. It is likely that woodblock printing developed a century or so before that.

Together these developments have come to be known as the Four Great Inventions, and are a point of pride in China today, where they are rightly seen as a reminder that Chinese civilization was considerably more advanced than the West in the centuries before the Renaissance.

# The Windmill

## First Invented: Persia
## Date: Ninth century AD

Watermills were used to grind grain in various areas around the world, including Greece, China, Scandinavia and the Roman Empire, from about the third century BC onwards. The most spectacular Roman version was a series of sixteen interlinked water mills on a descent at Barbégal in southern France. It was capable of generating over 30 horsepower and producing up to 27 tons of grain in a day (enough to feed over 10,000 people). The plentiful supply of rivers in the Roman Empire may be one of the reasons why their engineers didn't go on to experiment with wind power.

Hero of Alexandria had already used a wind-wheel to power a machine, while prayer wheels that spun in the wind were used in the Far East from about the fourth century AD. But it wasn't until late in the first millennium AD that this principle was applied to the mill – in Persia. Here wind was in more plentiful supply than running water. A tenth-century document suggests that the second caliph Umar, who ruled in the seventh century AD, constructed a windmill. However, the first authenticated reference comes from the Persian writer Istakhri, who described a windmill

The world's oldest windmills in Iran (formerly Persia). Their wooden blades rotate on a vertical axis between high walls of red clay and straw.

in which six to twelve sails made of matting or cloth rotated around a vertical axis, in a building with one side partially open to the wind. The structure looked somewhat like a revolving door.

This type of horizontal or *panemone* windmill spread around the Middle East, India and China over subsequent decades. The first references to vertical windmills (the type we would recognize more readily today) are found in European sources in the twelfth century. It is thought that this kind of windmill developed in Flanders first, but the first definite reference is from an 1185 description of a mill in the now-abandoned village of Weedley, adjacent to the Humber estuary in the Yorkshire Wolds.

## Underwater Diving

It has long been a dream of humankind to be able to explore deep beneath the ocean's surface. Of course, without breathing equipment, it is impossible to stay underwater for longer than a person can hold their breath.

From medieval romances we have a charming, albeit mythical, account of how Alexander the Great (356–323 BC) descended under the waves and travelled to the ocean floor in a glass diving bell. There he is said to have had an encounter with the King of the Fishes. Oddly enough, the first genuine account of breathing equipment being used for diving comes from the writings of Aristotle, who was Alexander's teacher as well as being remembered for his contributions to philosophy and science. He wrote that 'Just as divers are sometimes provided with instruments for respiration through which they can draw air from above the water, so also have elephants been furnished by nature with their lengthened nostril…', through which he explained they could breathe while walking underwater. It is not until about the eleventh century that there are accounts of more elaborate mechanisms being used, such as the pearl divers in the Arabian Gulf who strapped leather hoods filled with air around their chests to provide themselves with an air source on longer dives.

## CHAPTER 3

# MYSTERIES OF
# THE ANCIENTS

# Ancient Nanotechnology

### First Invented: Damascus
### Date: AD 300

We like to think of ourselves as being more intelligent and technologically advanced than our distant ancestors. But there are some examples of ancient technology which have defeated modern attempts at analysis, and which we remain unable to fully recreate. In particular there are several artefacts that rely on nanotechnology for their extraordinary qualities.

The people who created these artefacts were not relying on a true knowledge of the relevant science, but on trial and error in construction processes. Nonetheless, the evolution of their methods led them to a final product whose qualities depended on the manipulation of tiny, invisible particles (nanocomposites). Today we can use high-resolution microscopic analysis to uncover their molecular structure, but this is not enough to reveal the processes by which they were created, or to fully recreate them.

# Damascus Steel

One of the earliest examples of ancient nanotechnology is Damascus steel, which was manufactured in the Middle East between AD 300 and 1700. Weapons made from this steel were astonishingly hard, shatterproof and sharp, to a degree that was unmatched before the industrial revolution. Knowledge of how to make Damascus steel was lost after the eighteenth century. It is only recently that scientists have even been able to analyze samples of the metal in microscopic detail.

The crystallographer Peter Paufler, from the University of Dresden, found that the steel had a microstructure known as 'carbide nanotubes'. These are extremely hard tubes or wires made of carbon, which extend to the surface of the metal. The research identified numerous substances that were present in the metal in minute quantities, which had presumably caused the chemical reactions that created these nanotubes. These included bark of the *Cassia auriculata* tree, chromium, milkweed, vanadium, manganese, nickel, cobalt and some unidentified elements. It is possible that the crucial ingredient was one of these unidentified elements, and that the secret of the steel was lost once the miners had exhausted their supply.

So while Paufler's research has told us a great deal about

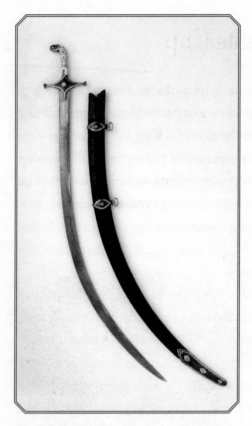

Weapons made with Damascus steel between 300 AD and 1700 AD were incredibly hard and sharp, but knowledge of how to make it was lost after the eighteenth century.

the production process which created Damascus steel, we still don't really know how it was made, and there is every chance we never will.

If you search for other images beware of modern knives sold under brand names like Damascus steel – it is probably best to stick to museum images which identify the sword or weapon as Damascus.

# The Lycurgus Cup

The Lycurgus Cup is a remarkable Roman artefact, a decorative glass cup from about AD 400 that is encased in a metal cage depicting the mythical King Lycurgus. The glass has a very peculiar property – it is dichroic, meaning that the glass shines in different colours according to the angle of the light. It is red when lit from behind, but green when lit from the front. While there are a few other fragments of

The Roman glass in this decorative cup from 400 AD is dichroic, meaning that the glass shines different colours according to the angle of the light.

Roman glass with this effect (using a variety of colours), this is the only complete surviving example.

In modern manufacture this effect can be achieved by layering glass with incredibly thin layers of metals or oxides. However, for a long time the process used to create the Lycurgus Cup remained a mystery. More recently, a few fragments of the glass were subjected to high-resolution microscopy. The strange effect turns out to be created by tiny traces of nanoparticles of silver and gold suspended within the glass. However, we have no way of knowing how the Roman artisans created this effect, particularly how they ensured that the silver and gold particles ended up in an even distribution throughout the molten glass before it set. While we can't reproduce the effect, it is at least possible to go and admire the beauty of the cup in the British Museum (although it is frequently on loan to other collections).

# Maya Blue

The pre-Columbian Mayan city Chichen Itza was the location of another strange historical discovery. Maya Blue, an azure pigment that was produced in approximately AD 800, is highly resistant to corrosion. Molecular analysis shows that it contains clay combined chemically with nanopores and indigo dye in such a way as to create a highly stable pigment.

The mechanism which makes Maya Blue work has been used by scientists to investigate new kinds of weatherproof and stable pigments. For instance, there has been research at the University of Turin into the possibility that the same method can be used for generating environmentally resistant paints in different colours, while the French National Centre for Scientific Research has explored a range of nanoporous materials in which organic dyes can be suspended.

Maya Blue is, therefore, a case where the inventiveness of the ancients has inspired new directions in our present-day understanding of nanotechnology. It may be inaccurate to suggest that the creators of these extraordinary artefacts fully understood the nanocomposites they were working with, but they are nonetheless continuing to give us food for thought today.

# The Baghdad Battery

## Created: The Sassanid Empire/modern-day Iraq
## Date: Third–sixth century BC

One of the many victims of the war in Iraq in 2003 was a mysterious ancient artefact known as the Baghdad Battery, or Parthian Battery, which was looted from the museum in Iraq's capital during the chaos of the invasion period. Reportedly discovered by archaeologists in the village of Khujut Rabu near the city in the 1930s, it consists of a terracotta jar 13 cm (5 in) in length, containing an iron rod encased in a copper cylinder. It was the first of a dozen similar finds, which all showed signs of corrosion on the inner surface of the vessels, revealing that they had contained an acidic agent.

The German archaeologist Wilhelm König came up with the extraordinary theory that this was an ancient battery, on the basis that an electrolyte solution such as vinegar or acidic fruit juice could have created an electric charge that was carried out of the lid of the jar via the copper cylinder and iron bar. König went on to speculate that the Sassanids may have known how to use this electric current to electroplate ornaments. In support of his theory, the Baghdad museum contained a variety of artefacts that had extremely thin plating.

Of course the fact that an ancient civilization knew how to generate electricity does not imply they fully understood the process, and the batteries would have lost power fairly rapidly due to their method of construction. But it is nonetheless fascinating to know that devices of this sort genuinely could have generated a modest amount of electricity, as modern reconstructions have demonstrated. Other experts have, however, disputed König's suggestion that the purpose was electroplating, noting that the objects in the museum could easily have been created using non-electric plating techniques that were known in the period. And there is no evidence for the batteries having been joined together in a series, which would probably have been necessary to produce the higher current required by electroplating.

One alternative suggestion is that the mild electric current from the battery could have been exploited for therapeutic purposes. For instance, it might have been applied to the soles of the feet in an early form of electrotherapy.

Whatever the explanation is, the batteries were yet another demonstration that ancient civilizations had some technology that was far more sophisticated than we tend to imagine. It is a shame that their disappearance may mean we no longer have the chance to carry out future tests to explore their origins and purpose.

# Unbreakable Glass

## First discovered: Rome
## Date: First century AD

There is a story, which may be apocryphal, that suggests the Romans discovered unbreakable glass many centuries before the reinforced glass we know today. It comes from the reign of the Emperor Tiberius, who is remembered for his military successes but also for being a rather gloomy and reclusive individual. Apparently, a well-known inventor of the day visited the emperor and informed him that he had made a remarkable discovery – he had learned the secret of making glass that couldn't be broken under any circumstances.

Tiberius asked the inventor whether anyone knew the secret other than the two of them who were in the room together. The inventor confirmed that the emperor was the first to hear about it. At that, Tiberius called his guards into the room and had them drag the poor inventor away to be executed. Tiberius's problem was that he didn't want to go down in history as the emperor who had put all his glassmakers out of business.

# Roman Concrete

### First Used: Rome
### Date: First Century

When it comes to the question of 'What have the Romans ever done for us?', one answer in the future may well be, 'They taught us how to make better concrete'. For a long time it has been known that Roman concrete is remarkably durable, especially when it is used underwater in sea walls, breakers and piers. For instance, the harbour at Caesarea in Israel, which was built late in the first century BC, is made of astonishingly hard concrete that has maintained its shape well through the centuries.

Until recently the secrets of how Roman concrete was made were a mystery, although it was known that the Romans were proud of its construction. Pliny the Elder wrote proudly in the first century BC that it was 'impregnable to the waves and every day stronger'. The basic ingredients of the concrete are given in the writings of Vitruvius in the previous century. For structural mortars he prescribes the use of *pozzolana* (volcanic sands from Pozzuoli, where the sand is brownish-yellow, or from the coast near Naples, where it is reddish-brown), which should be mixed with lime (calcium oxide) in a ratio of 3:1.

Casearea Maritima was a town built in the time of King Herod and its concrete structures are still clearly visible.

However, analysis of samples also showed that the concrete contains the mineral aluminous tobermorite, which the Romans wouldn't seem to have had the technology to manufacture. American geologist Marie Jackson is the co-author of a study into Roman structures, including their concrete. Their report describes their analysis of the concrete from Roman piers and harbours. Remarkably, they found that the aluminous tobermorite actually formed within the concrete, as the lime and volcanic ash reacted with the seawater and generated heat.

In addition, the team found that tobermorite was

continuing to grow within the concrete, along with phillipsite, another crystalline mineral. As seawater seeped into the concrete, it continued to react with crystals in the volcanic ash. This meant that, rather than cracks developing as tends to happen in modern concrete, any slight fissure would actually be continually reinforced by the production of tobermorite and phillipsite.

This suggests that we could learn from the Romans and incorporate similar chemical mixtures into future underwater concrete structures, and thus make them as wonderfully long-lasting as their classical precursors.

# The Assyrian Nimrud Lens

## Discovered: Modern-day Iraq
## Date: 750 BC

One of the mysteries of the ancient Assyrian civilization (which lasted from about 2000 BC to about 600 BC) is that they seem to have had a remarkable degree of astronomical knowledge – in particular, they depicted Saturn as a god surrounded by a ring of serpents. Some scholars have conjectured that this might mean they had somehow managed to observe the rings of Saturn, but this is surely

The Nimrud lens, c. eighth century BC, in the British Museum, London.

impossible without a telescope? Telescopes were not invented until the seventeenth century (see page 178).

It is possible that this mystery can be explained by the 'Nimrud lens', a piece of transparent rock crystal that was found by Sir John Layard in the Assyrian palace of Nimrud. Dated to about the eighth century BC, it is an oval shape and has been deliberately ground to give it a fairly smooth surface. The result is a kind of lens that can magnify objects by up to three times their size, although the focus is not very clear. As with other ancient lenses, it is generally believed to have been used as an ornament, as a burning-glass for starting fires, or possibly as a short-range magnifying glass.

Sir John Layard's suggestion was that the glass may have been used by Assyrian craftsmen to make minute, detailed carvings, as are found on other artefacts from the period. By contrast, Italian scientist Giovanni Pettinato believes that the lens could have been used in sequence with other lenses to create a telescope, which would have been somewhat out of focus, but possibly strong enough to give the Assyrians additional information about the solar system and the stars – possibly even clear enough to see that Saturn was surrounded by some kind of ring, for instance.

This is a fairly revolutionary suggestion, and one that is rejected by many historians. While it is known that there are many ancient lenses – some of the earliest come from 2500 BC (in ancient Egypt) and 1500 BC (in Knossos, Crete) – the scholarly consensus is that these were curiosities or household tools and that the method of aligning them to create a telescope hadn't been discovered by the ancients. However, as many of the remarkable discoveries mentioned in this book show, we sometimes give our ancient ancestors less credit than we should, so we should at least entertain the possibility that Assyrians really were millennia ahead of us in using a telescope to observe the heavens.

# The Norse Sunstone

First Used: Iceland
Date: Thirteenth century

There are a variety of references in Icelandic texts to the sunstone, a marvellous crystal that could magically reveal the direction of the sun even in a cloudy sky or after it had set behind the horizon. For instance, the Icelandic sagas tell of the Viking king, Olaf, asking a servant, Sigur, to point to the location of the sun on a snowy day. To check the answer, 'the King made them fetch the solar stone and held it up and saw where light radiated from the stone and thus directly verified Sigur's prediction'. Such a crystal would provide an extremely useful navigational device, particularly in the northern waters in which the Vikings sailed. We know that they were remarkable navigators, as proven by the fact that they found their way to Greenland and the coast of North America centuries before Christopher Columbus. But such a sunstone sounds too good to be true, so most historians have dismissed the story as a myth. At least, they did until recently.

The discovery of a cloudy crystal in an Elizabethan shipwreck off the Channel Island of Alderney has helped to increase scientists' faith that the sunstone might be a real phenomenon. It was about the size of a pack of playing cards, was found next to other navigational devices, and was made of Icelandic spar, which might confirm that not only was the sunstone real, but that the knowledge of how to use it was still retained by Tudor sailors.

The explanation for this lies in the speculation of Danish archaeologist Thorkild Ramskou that the sunstone might be made of a mineral such as Iceland spar or cordierite that polarizes light. When a team led by Guy Ropars looked into this theory, they found that a piece of Icelandic spar could indeed be used to find the hidden sun's location to within a few degrees. You move the stone in front of your eyes and look at the pattern of the light. If a dot is placed on the upper surface, when viewed from below two dots are perceived, at different degrees of brightness. This is because the sun's rays are being depolarized and being shifted along different axes. If you rotate the crystal until the two points are equally bright, then you are pointing it directly at the sun.

It has to be said that attempts by other scientists to replicate the results have been inconclusive, but the work of Ropars's team, along with the discovery of the possible example in the wreck of the Tudor ship, does give us reason to suspect that the sunstone may not be a myth after all, but a remarkable piece of ancient technology which had been lost to the descendants of its original owners.

# The Antikythera Mechanism

## First Created: Greece
## Date: Second century BC

The island of Antikythera lies between Crete and Greece (the name means 'opposite Kythera', which is a large island adjacent to it). In 1900, a Greek diver found a wreck just off the coast of the island. Presumed to be a Roman vessel from the first century BC, it contained many remarkable artefacts that dated back as far as the fourth century BC. It's possible that the boat was carrying booty from the sacking of Athens in 87–86 BC, or that the objects were being taken to Rome to be presented to Julius Caesar.

Many of the objects retrieved from the wreck were quickly identified, but the most mysterious and fascinating artefact – the Antikythera mechanism – puzzled historians for decades. It was a damaged remnant of a wooden box (about the size of a shoebox), which contained thirty interlinked gears and levers. The famous physicist Richard Feynman described it as 'so entirely different and strange that it is nearly impossible … it is some kind of machine with gear trains, very much like the inside of a modern wind-up alarm clock'.

It befuddled archaeologists and was more or less ignored

A model of the Antikythera machine.

until 1951, when the historian Derek de Solla Price became intrigued by it. After two decades of work, he published some preliminary thoughts, but he hadn't fully understood the device before his death in 1983. However, the attention of academics was starting to focus on a mention in the writings of Cicero of a mechanical planetarium known as the 'sphere of Archimedes' that showed how the celestial objects in the solar system moved relative to the Earth. The inscriptions on the device, in Koine Greek, seemed to back up this theory, as did the fact that the dials and rings were inscribed with Greek zodiac signs and calendar days.

In the past few decades it has been established that the mechanism is indeed a device for tracking the lunar calendar, and predicting eclipses and the position and phase of the Moon. In addition it shows the seasons, and marks festivals such as the Olympics. One of its most remarkable properties was that it could calculate the Moon's period to a high degree of accuracy and model its elliptical orbit. As a result it has been called 'the world's first analog computer', which might seem hyperbolic, but given that mechanical aids such as slide rules or tide predictors were in use before digital computers, and *they* can be described as analog computers, it is pretty accurate.

The remaining mysteries about the mechanism concern who owned it and who created it. One suggestion for the creator is Hipparchus, the second-century BC astronomer and mathematician who is remembered for compiling the first trigonometric table and for his rather accurate estimate of the distance from the Earth to the moon. Cicero suggests that Posidonius, who followed Hipparchus as the head of the school on Rhodes, built a planetary device, which might mean that the two men collaborated on its design.

That mystery may never be solved, and we can't know whether this is the one surviving example of a machine that other people were also able to build. However, the mechanism (and modern recreations of it) shows us how remarkably sophisticated some Greeks of the period had become.

## CHAPTER 4

# MILITARY
# INVENTIONS

# Archimedes' Heat Ray

## First Invented: Syracuse, Ancient Greece
## Date: 214–212 BC

The polymath Archimedes of Syracuse (287–212 BC) was one of the pre-eminent philosophers and scientists of antiquity. As well as explaining the principles of how levers worked, he invented numerous extraordinary machines and devices, including the Archimedes Screw (which is still used today to pump liquids), and the first compound pulley system (see page 64).

He is also famous for the war machines and catapults he designed to help protect his native city from attack by the Romans. One example, known as the 'ship shaker', was an enormous crane to which a metal grappling hook was attached. When it was swung out over the water, the claw could be dropped onto the ship, crushing it and causing huge damage. A 2005 television documentary, *Superweapons of the Ancient World,* tested the device using only materials available in the ancient world, and found that it was indeed a viable invention.

The most spectacular claim about Archimedes' weapons is that he created a 'heat ray' (or death ray) that was capable of focusing the power of the sun into a powerful beam, capable of igniting wooden enemy ships. Accounts of how this worked vary – one version has Archimedes using concave mirrors to focus the lightbeams. In another version he asks the soldiers of the city to use their polished bronze shields.

There is some debate about whether the machine

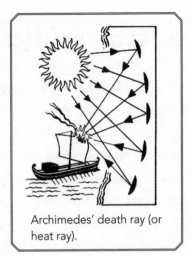

Archimedes' death ray (or heat ray).

ever existed, or is even feasible. In 1973 the Greek scientist Ioannis Sakkas built a version of the death ray using 70 mirrors coated with copper and aimed it at a plywood model warship. The model rapidly caught fire and burned to ashes. However, it was only 50 metres (164 feet) from the weapon, which seems too close to make the experiment valid – at this distance, a volley of burning arrows would have been a much easier option. Also, plywood burns fairly easily compared with woods such as cedar, which would have been a more likely material for a Roman battleship.

More recently the American TV programme *Mythbusters* has made several attempts at revisiting the experiment, by challenging viewers to create a workable death ray. In 2005, students at the Massachusetts Institute of Technology claimed partial success, as they managed to set a fake boat on fire from a distance of 30 metres (98 feet). They replicated the experiment from a greater distance for the show, but it was deemed to have failed – while the target partially ignited, it wasn't an efficient or fast enough process to match up to the fearful weapon described in the stories of Archimedes. The debate will undoubtedly continue until, and unless, someone can build a truly successful working model of the ancient death ray.

# The Ancient Machine Gun

## First Invented: Greece
## Date: Third century BC

The machine gun is a weapon that we would think of as being undoubtedly modern. However, there was an ancient Greek version called the *polybolos*, for which we have a technical specification in the artillery manual of Philon, late in the third century BC. Invented in the arsenal at Rhodes by Dionysius of Alexandria, it was a magazine-fed bolt-shooter, which could fire repeatedly. It was regulated by a chain drive that ran over rotating sprockets. This drew back the bowstring, which was in turn powered by a rotary feeder, then pressed the trigger at exactly the right moment. It was

Ancient military artillery, all variations of catapults,
c. third century BC.

loaded with bolts, which it could fire at a remarkable speed.

We have a fairly clear idea of how effective the *polybolos* could have been. A German military engineer (whose name was Schramm) constructed a working model for Kaiser Wilhelm in the late nineteenth century. He used a bicycle chain to recreate the original contraption. Curiously, the *polybolos* was criticized at this time for its accuracy – one contemporary writer commented that the lack of variation in where the bolts landed meant that it concentrated fire too heavily on one target. Schramm's model was extremely accurate. The second bolt he fired actually split the first one that had landed in the target ahead in half, making him something of a teutonic Robin Hood.

# A Very Brief History of Early Weaponry

Humankind has constructed weapons from prehistoric times. The earliest weapons were clubs, stone axes and spears, all of which were being used at between 100,000 and 500,000 years ago.

- The oldest known spear, made of spruce wood, is the Clacton Spear, the tip of a wooden weapon discovered in 1911 in Clacton-on-Sea, England.

- One current theory suggests that the age of the spear was a relatively peaceful period in human history, as a single man armed with a spear was still dangerous to any would-be attackers.

- Primitive slings were also probably used in the paleolithic period as a way of throwing rocks to hunt game.

- Throwing spears were a significant step, as they allowed humans to kill from a greater distance. The oldest known are the Schöningen spears excavated from an ancient mine in Germany, and dated at approximately 350–

400,000 years old, although some experts have argued that these may have been thrusting spears.

- Early aborigines were armed with spears, war boomerangs (which weren't designed to come back, but to injure their targets) and curved sticks for fending off attackers.

- The bow and arrow was probably first invented about 50–65,000 years ago in Africa. The oldest intact bow comes from Denmark about 10,000 years ago, but arrowheads and tips have been found from much earlier cultures.

- The bow and arrow led to the earliest organized warfare, as it enabled attackers to use the same weapon repeatedly. The oldest evidence of large-scale warfare comes from Cemetery 117 – a site in Sudan where at least fifty-nine bodies were buried about 14,000 years ago.

- Biological warfare dates to at least 1700 BC, when the Sumerians used a variety of methods of intentionally spreading contagion as a weapon.

- Poison arrows may have been in use by about 1200 BC in the Trojan War – toxic weapons feature in Homer's *Iliad*. By the sixth century BC the Scythians (from modern-day Iran) were notorious for their virulent poisoned arrows.

# The Crossbow

## First Invented: China
## Date: c. Sixth century BC

The crossbow is a mechanical application of the bow-and-arrow principle. It generally consists of a horizontal bow (known as a 'prod'), which is mounted on a stock. The projectiles that it fires are called bolts or quarrels. Crossbows were another significant step in people's ability to wage war. While archery was a highly skilled craft that generally had to be learned from childhood by dedicated archers, the crossbow could be mastered by any soldier or new recruit with a few weeks' training. This enabled far more armies to get up to fighting condition within a short amount of time.

The earliest definite evidence we have of the crossbow comes from the sixth century BC in ancient China and the neighbouring areas. A fourth-century BC text mentions a giant crossbow being used in the sixth or fifth century BC, while Sun Tzu's classic text on military tactics, *The Art of War*, which dates to 500–300 BC, mentions the crossbow several times.

When it comes to artefacts, bronze crossbow bolts that date to the mid-fifth century BC have been discovered in burial sites around China, while crossbow stocks small enough to

A Roman *ballista* crossbow.

be handheld have been found at a dig in Qufu, Shandong, dating to the sixth century BC. A more controversial question is when repeating crossbows were first used. These are crossbows that can rapidly fire multiple bolts. There is some suggestion these might date to earlier, but they are generally credited to the famous military adviser Zhuge Liang (AD 181–234). His version, which would be deadly when used by massed ranks of soldiers, could fire two to three bolts at once. It had a magazine loaded with bolts over the bow, and a lever-driven mechanism to replenish the bolts. The weapons of this period had a range of about 100 metres (330 feet). In the medieval period, the Chinese also developed a 12-round repeater crossbow that continued to be used until the nineteenth century, and has been compared to the machine gun in terms of its destructive capacity.

# The Catapult

## First Invented: Greece
## Date: Fourth century BC

Following the invention of the crossbow, it was only natural that people would look for ways to extend the principle to large weapons. As crossbows grew, the larger version became too unwieldy to be handheld, and the catapult – the first form of artillery weapon – was born. The crossbow was in use in Greece in a similar period to the first Chinese versions. Diodorus Siculus, a Greek historian, gives us the first recorded mention of a larger, mechanical arrow-firing catapult (known as a *ballista*) in 399 BC. (There is also a much earlier, but disputed, mention in the Bible of a weapon that might be an early catapult, which would date to the eighth century BC. In II Chronicles 26:15 it is said that King Uzziah 'made in Jerusalem engines, invented by cunning men, to be on the towers and upon the bulwarks, to shoot arrows and great stones withal.')

A *ballista* really is just a giant crossbow, which is freestanding on its own stand, and in which extra force is applied to the reinforced string of the bow by using a winch to hoist it back into a firing position. It could be used to fire large arrows, sharpened sticks or

A *trebuchet*.

blunt beams, from a groove in the central beam.

Also in the fourth century, the ancient Romans invented the first type of catapult that resembles our modern idea of what they look like – the *mangonel*. Instead of the bowstring directly propelling the projectile, the design was adapted so that a string hauled an upright hinged beam back towards the horizontal, where it was momentarily immobilized with a powerful catch. A bucket on the end of the beam was filled with the projectile, then the catch was released. When the beam swung back into a vertical position it collided with a central crossbeam, which stopped its motion, and the momentum carried the projectile forwards in a fast, flat trajectory. The *mangonel* could be used to attack the walls of a fort, or to hurl projectiles into an opposing line of adversaries.

The final development in the story of giant catapults was a combination of the catapult with an ancient weapon – the sling – to create the *trebuchet*. A wooden beam, which could be more than 15 metres (50 feet) in length, was hinged on a fulcrum near to one end of the beam. The short end was weighted with a heavy object, which initially rested on the ground. The long end of the beam was pulled towards the ground (either manually or with a winch), and a sling on the end of it was loaded with projectiles. *Then* the beam was released, the heavy weight fell to the ground, and the sling released the projectiles. The *trebuchet* was mainly used as a siege weapon, as it could fire extremely heavy objects high into the air and over the walls of a fort or castle.

The *trebuchet* and *mangonel* could be used to fire a far wider range of missiles than the *ballista*. Over the centuries they were used to fire quicklime, burning tar and sand, animal dung, dead animals and humans (or parts of their bodies), and diseased bodies (such as plague victims) in order to spread fear and chaos among adversaries as well as to destroy the target. The *trebuchet* in particular was one of the most formidable weapons of the pre-gunpowder era, and continued to be used throughout the medieval period in Europe.

# The Warship

## First Used: The Mediterranean
## Date: Second millennium BC

Boats and canoes were probably used in prehistory to transport fighters and for close combat. However, for the first true warships we need to look to the maritime cultures of the Mediterranean and Middle East. The Phoenician culture of the eastern and southern Mediterranean coasts was especially known for its seagoing skill and shipbuilding, and is sometimes credited with the earliest warships. The truth is probably a bit more complicated. As ships became larger and larger, they would have been used increasingly in warfare as they became more valuable prizes to attack. And larger boats were commonly used in the second millennium BC for carrying out violent raids on other areas of the coast (a piratical activity that the later Greek historian Thucydides described as having been widespread and virtually without stigma in the ancient period.) This made it increasingly crucial to work out forms of defence as well as offence on the seas.

The most commonly used ships for warfare were galleys, powered by rowers. Speed was of the essence, as the main military tactics relied on outmanoeuvring your opponent.

The technical names for galleys referred to the number of oars (or the number of banks of rowers) that powered them. Early Greek galleys with a single bank of thirty oars are known as *triaconters*, for instance, while any single-banked galley can be referred to as a *monoreme*, a double-banked one as a *bireme*, and a triple-banked one as a *trireme*. The banks were necessary to create more powerful vessels, because ships with room for more than about thirty rowers per side were too large for nimble manoeuvres.

The earliest warfare involved attempting to board your opponents' vessel and defeating them in combat. However, sometime early on in the first millennium BC, things became more complicated with the invention of the ram. A projecting beam at the base of the bow of the ship was covered in metal, and could be used to punch a significant hole in the enemy's craft. Ships of this ilk were certainly being used by the Phoenicians by about 750 BC.

The development of artillery such as catapults, from the fourth century BC onwards, once again transformed the nature of battleships. Rather than ram an opponent face on, the galleys would attempt to take up a position from which they could do most damage with their long-range barrage, and the ship with a greater range of weaponry held a crucial advantage.

The first Punic War in the third century BC was fought between the growing empire of Rome and Carthage –

the city-based empire that had grown from its origins as a Phoenician colony. In the early naval exchanges, the Carthaginians had the upper hand. As a result, Rome built a new fleet of warships that were based on a captured Carthaginian *quinquereme*. Their main modification was the *corvus*, a spiked platform that could be lowered onto an adjacent enemy ship so that the soldiers could storm across. This allowed the Romans to apply something closer to their land-based military tactics to the remaining battles of the war, and was a factor in their eventual victory. From this point onwards the Romans were the dominant naval power in the Mediterranean and beyond, but it is notable that most of their shipbuilding knowledge was acquired in the first place from the earlier cultures that they had supplanted.

---

# Greek Fire

### First Invented: The Byzantine Empire
### Date: Seventh century AD

One of the genuine mysteries of antiquity is Greek fire, a mysterious Byzantine chemical weapon which could even burn in water. The secret of how it was made remains unsolved today.

The earliest incendiary weapons were simple torches and burning arrows, which were used (among other places) in China and the Assyrian Empire in the first millennium BC, and there are reports of burning liquids (possibly related to oil) being used from the early part of the same period. But Greek fire became famous because it was such a dramatic escalation of incendiary warfare. The best-known use of it was in AD 678, during the defeat of a rampant Saracen fleet that set out to conquer Constantinople, the capital of the Byzantine Empire. (Constantinople is now called Istanbul. At the time, it was the main city of the eastern Roman Empire. The Saracens were Islamic Arabs of the Umayyad Caliphate.)

The Saracens' advance had been a successful one and they were confident of victory in the forthcoming naval battle, which would have given them control of the seas around Constantinople. The advancing Byzantine navy was known to be numerically inferior. However, the Byzantines had a secret weapon concealed behind bronze tubes that jutted out from their ships' prows. As they came closer, these tubes suddenly emitted streams of liquid that were ignited and turned into pure fire. Greek fire, as it became known, stuck to every surface it touched, set fire to anything it came into contact with, and even burned in the water, meaning that those poor sailors who were ignited in the initial bursts of battle couldn't extinguish the flames after they leapt into the water. The Saracen fleet was partially demolished and the remnants fled in terror, to spread the story of Greek fire around the Mediterranean and beyond.

After that incident, Greek fire was used relatively rarely, partly because the legend of its horrific effects was sufficient to deter most fleets from attacking the Byzantine fleet. However, it was used to defend the walls of Constantinople on a number of occasions, and also fired as a projectile in the form of a ball of cloth soaked in the liquid and fired from a catapult. It had its limitations. When fired from a ship it had a limited range, as it had to be pumped from a vat through the tubes (which also made it impossible to use into a headwind), while the catapult method involved smaller quantities of fire. However, it remained a fearsome weapon that had a huge deterrent effect on enemies.

The substance responsible had been developed by the Greek architect Kallinikos over the previous decades, and was a jealously guarded secret, to the point that only his family and the Byzantine emperors knew the recipe. It was passed down from one emperor to the next until the secret was finally lost.

Some modern experts have speculated that the main ingredient may have been calcium phosphide, which can be made by heating lime, bones and charcoal. When in contact with water, it releases phosphine, a substance that ignites spontaneously. Other possible ingredients include tree resin (which would explain its stickiness), some derivative of crude oil (as in the manufacture of napalm), and quicklime. But while experiments with these substances have come close to replicating the historical reports of Greek fire, the exact recipe will remain a secret forever.

## One That Got Away

Sometimes the ingenuity of ancient inventors ran ahead of the technology available, with the result that their ideas weren't realized. The Alexandrian engineer Ctesibius carried out many experiments with torsion springs (whose power comes from metal or another substance being twisted). In particular he wanted to build a catapult that used hinged levers whose movement was driven by bronze compression springs. However, the metals available to Ctesibius were not sufficiently springy for his idea to work. It wouldn't be until about AD 1400 that the forged steel created by Italian blacksmiths was robust enough to achieve this. Similarly, a contemporary design for a pneumatic catapult, which relied on compressed air contained in cylinders, was never built. The type of pressure resistance and detailed mechanisms that this would have required wouldn't be feasible until the industrial age, when precision gauges and machine tools were invented.

# Poison Gas

## First Used: China
## Date: Fourth century BC

The earliest use of poison gas was possibly in ancient Egypt, where gas was used to control fleas. However, the earliest record of its use in warfare comes from China in the fourth century BC. When a city was besieged, one strategy that attackers could use was to dig tunnels beneath the city walls, both to undermine the structure and as an invasion route. Defenders learned to tap into the tunnels and pump gas in using bellows. The substances used included burning mustard balls and the artemisia plant, both of which produce a noxious gas when burned. By the time of the Roman Empire, a wider variety of poisons were available. One standard Roman strategy was to poison the wells of cities they were besieging.

The earliest physical evidence of poison gas comes from AD 256 at a Roman fort in the abandoned city of Dura-Europos in modern-day Syria. The corpses of twenty Roman soldiers were found under the ramparts. The story goes that Persian attackers attempted to burrow under the walls of the city. In response the Romans dug their own tunnels for the purpose of defence or escape.

The Persians responded in turn by pumping fumes from a burning mixture of sulphur crystals and bitumen through holes into the tunnels. They continued to do this 'until the screaming stopped.' After the fall of the city, the Persians dug the corpses out of the tunnel and concealed them under the walls.

Another fascinating story from history comes from Themiscrya, a Greek outpost that the Romans were attacking. In this case, the Greeks didn't have access to poison, but they did have beehives, which they succeeded in lowering into the tunnels. The tunnelling Romans were driven back by a barrage of angry bees.

# Armour and Tanks

## Invented: Sumerian Civilization
## Date Third millennium BC

The earliest evidence we have of armour comes from the first half of the third millennium BC, from the Sumerian city of Ur. A decorated box, known as the 'Standard of Ur', is decorated with an image of a marching army wearing helmets, tunics and heavy cloaks. At this stage of history leather was used for armour, which provided some protection but could be pierced by arrows and lances. The Chinese partially solved this problem over time by applying lacquer to their leather armour, which gave it a much harder surface. Twelve magnificent lacquered armour suits were recovered from the tomb of the Marquis Yi of Zeng from 433 BC.

The first metal armour came in the form of helmets. For instance, the same archaeological dig that produced the Standard of Ur also uncovered a metal helmet made of beaten gold and silver (suggesting that it was ceremonial rather than for use in battle). By the fifteenth century BC metal armour was in use – a tomb at Dendra in southern Greece contained a nobleman in a full suit of metal armour. In the same period, the Chinese addressed the problem of the weight of metal armour by making a hybrid metal and leather version. Bronze plates mounted on leather also made these suits of armour relatively flexible.

Some horses and even elephants also went into battle in armour in ancient history. For instance, one of the defining moments in Alexander the Great's campaigns in India came at the hands of King Porus of the Paurava kingdom in the area around the river Jhelum (in the modern-day Punjab province of Pakistan). Alexander's army were taken aback by the ferocity of their opponents, and had particular problems dealing with elephants clad in leather armour. While they won the battle, it was at great expense, and rumours of kingdoms further east with even larger troops of war elephants deterred Alexander from continuing to invade any further into the subcontinent.

By the first millennium BC in the Assyrian empire, the idea of armoured vehicles had been put to use in the form of 'siege engines', which provided cover for battering rams attacking city gates. A dense wickerwork carriage on four to six wheels contained the ram and the soldiers who operated it, while towers on the top of the carriage were used by archers to attack the defenders on the walls. In some versions, there was even a water tower to extinguish incendiary missiles (which were the greatest danger to the operation).

The biggest step towards what we would now recognize as a tank came in China. An iron-plated armoured car was a key weapon in the twelfth-century AD battles against the Tatars, who relied heavily on horse-mounted cavalry charges in their battle tactics. Early tanks were unwieldy, but could convey a huge advantage. During the religious conflicts of the fifteenth century, there was a peasant uprising led by Jan Zizka against the mighty German Imperial army in Bohemia. Given limited resources and numerical limitations, Zizka had to be creative. He had wagons covered in sheets of iron light enough to be pulled by a team of horses. Inside the tanks his peasant soldiers were armed with spiked iron bars, axes, crossbows and simple handguns. Even though they had 25,000 men against Imperial forces of up to 200,000 they had several notable victories before the rebellion finally petered out due to sheer exhaustion.

## The Parachute

The first modern design for a parachute was sketched by Leonardo da Vinci in about 1485. However, the first use of atmospheric drag to slow the movement of a falling person was in ancient China. In *Historical Records* by the second-century BC historian Si Ma Chian, there is a story about the emperor Shun (who is said to have lived in the twenty-third century BC) in which he is cornered by his enemies on a roof, but escapes by holding two large bamboo hats in his hands in order to fall safely to the ground. While that story may be mythological, we do know that circus performers in China used similar methods to fall gracefully from significant heights from 200 BC onwards.

## CHAPTER 5

# MEDICAL KNOWLEDGE

# Surgery

## First Practised: Iraq
## Date: 60,000–30,000 BC

Ancient civilizations practised a surprising variety of surgical procedures, including suturing or cauterizing wounds and amputating limbs. Evidence of this knowledge from about 5,000 years ago is fairly widespread, and includes Stone Age needles in Africa that seem to have been used for sewing-up cuts. (For the same purpose, some tribes in both India and South America sealed wounds from minor injuries by placing termites or scarabs around the edges of the cut, then detaching the heads from the bodies mid-bite, leaving the locked jaws to seal the flesh.)

Trepanation (drilling a hole in the skull) is one of the earliest known surgical procedures practised by humans. A burial site from about 6500 BC contained 120 skulls, of which 40 had the characteristic burr holes. In some societies it was believed that this practice would release evil demons or spirits. However, it was also used for the more legitimate aim of releasing blood from the area around the brain after cranial injuries, making it the first example of brain surgery in history.

Amputation is also an ancient practice. A skeleton from about 7,000 years ago that was found in a grave at

Buthiers-Boulancourt, south of Paris, had had its left forearm intentionally amputated, probably with a flint knife. Scientists have examined and tested the remains speculated that the patient might have been anaesthetised using pain-killing plants such as Datura, a poisonous vespertine with hallucinogenic properties, while herbs such as sage could have been used to clean the wound (which remarkably showed no signs of having become infected).

However, bones found in the Shanidar cave in Iraq suggest that surgery may have been practised even earlier in history by our Neanderthal cousins. The skeleton known as Shanidar-1 (nicknamed 'Nandy' by the archaeologists who found him) had numerous injuries – including a severe blow to the face which would have caused partial blindness, blockages in the ear canals that would have led to a degree of deafness, and a withered right arm with several healed fractures and a missing lower arm and hand. This discovery has already transformed our understanding of Neanderthal culture – all of these injuries were sustained a long time before 'Nandy' died, suggesting that sick or elderly members of their tribes were looked after. In addition, it has been speculated that the lower arm and hand were deliberately amputated, given the way in which it had healed after the loss. The skeleton's date is between 60,000 BC and 30,000 BC, so if this was a surgical operation, it would be the earliest known within any of the human or near-human species.

# Plastic Surgery

## First Invented: India
## Date: Sixth century BC

You would probably imagine that plastic surgery is a fairly recent medical advance. However, there is evidence of it being performed at least 2,600 years ago. An Egyptian manuscript, whose contents were copied from a text from around 2500 to 3000 BC, contains instructions that have been interpreted as a description of reconstructive surgery for a broken nose.

The later evidence from India is, however, much clearer. About 3,000 years ago there was a local tradition of ear stretching, in which children's ears were pierced and then progressively stretched to make the holes bigger. It may have been this tradition that inspired the idea that flesh and skin could be manipulated. Either way, in the text known as the *Sushruta Samhita* (from the sixth century BC), there is a detailed description of surgical procedures including nose reconstruction (or 'rhinoplasty'):

> *The portion of the nose to be covered should be first measured with a leaf. Then a piece of skin of the required size should be dissected from the living skin of the cheek, and turned back to cover the nose, keeping a small*

*pedicle attached to the cheek. The part of the nose to which the skin is to be attached should be made raw by cutting the nasal stump with a knife. The physician then should place the skin on the nose and stitch the two parts swiftly, keeping the skin properly elevated by inserting two tubes of eranda [the castor-oil plant] in the position of the nostrils, so that the new nose gets proper shape.*

Finally the writer suggests applying licorice, red sandalwood and barberry plant, before covering the affected area with cotton and clean sesame oil. The *Sushruta Samhita* also describes the use of cheek flaps in reconstructing ear lobes, using wine as anesthesia, and using leeches to keep wounds free of blood clots. The book became one of the founding texts of the Ayurveda, the traditional medical system of India, and contains instructions for over 300 surgical procedures. Arabic translations preserved this knowledge and eventually spread it to the West during the Renaissance period.

The influence of Indian medicine on plastic surgery in the West didn't end there. In 1794, an article in the *Gentleman's Magazine of London* recounted the plastic surgery that British soldiers had seen being used to repair the nose of a Maratha cart-driver who had been injured in battle. The account inspired doctors such as Joseph Constantine Carpue to travel to India to learn more about their remarkable tradition of plastic surgery.

# A Very Brief History of Anatomy

Some of the earliest explorations of the human body came when the bodies of victims of sacrifice or execution were examined.

- A papyrus found in 1600 BC demonstrates that the ancient Egyptians already recognized the heart, spleen, kidneys, uterus and bladder, and knew that blood vessels were connected to the heart.

- The early Greek scientist Alcmaeon (the sixth century BC) was the first to practise anatomical science by dissecting animals. He also identified the optic nerve.

- In the fifth century, Empedocles argued that the heart was the main organ of the circulation of blood and of 'breath' or 'soul'.

- By the time of Hippocrates (460–370 BC) there was a better comprehension of the role of the kidneys and of how the heart worked.

- During the fourth century BC, Praxagoras identified the fact that arteries and veins are distinct from one another.

- Ptolemy I Soter, ruler of Egypt during the third century BC, gave permission for a school of anatomy to be set up in Alexandria, where dead bodies would be dissected. (This was previously a taboo in most cultures.)

- The anatomist Herophilis in Alexandria argued against Aristotle's theory that the seat of consciousness was the heart, and demonstrated that it was actually in the brain.

- Most medieval knowledge of anatomy came from the work of Galen, a second-century scientist who studied animals as well as gladiators who had been badly wounded. He produced twenty-two volumes of detailed anatomical knowledge.

- Galen argued correctly that arteries carried blood (not air as was commonly believed), but wrongly thought that blood ebbed and flowed from the heart rather than circulating.

- Emperor Frederick II (1194–1250) made it mandatory for medical students in the new Italian universities to learn about human anatomy and surgery. The anatomist Mondino de Luzzi was reputedly the first to dissect human bodies as a demonstration during lectures.

# Socialized Medicine

## First Invented: Greece
## Date: Fifth century BC

The funding of doctors and hospitals is a controversial political issue in many countries in the modern day. We think of social democracy, and the provision of healthcare by the state, as a relatively recent notion. However, it is an idea that goes back to the origins of modern medicine. In the fourth century BC, the Hippocratic Precepts, which set out the Hippocratic ideal of how doctors should operate, advised 'Sometimes give your services for nothing, calling to mind a previous benefaction of present satisfaction. And if there be an opportunity of serving one who is a stranger in financial straits, give full assistance to all such.'

By the time of the Roman Empire, the rich paid for doctors but a public physician was also provided for the ordinary citizens by the local town councils. However, this was not always medical care of the highest quality. In 220 BC Rome's first public physician was the Greek doctor Archagathus, whose preferred cures tended to involve the cauterizing iron and the surgical knife. His nickname was 'the Butcher', and the politician Cato the Elder was convinced he was part of a Greek conspiracy to murder all Romans.

The Chinese took the idea of socialized medicine further, by using a system of publicly employed doctors who were paid for by the central government. Initially set up in major cities in the second century BC, this system was extended to the whole country by the first century AD. Full training was also provided. By the fifth century, most major cities had a dedicated medical college that granted degrees in medicine.

# False Teeth

### First Invented: The Etruscan Civilization
### Date: 750 BC

The Etruscans were the ancestors of the Romans, a wealthy civilization that occupied the central area of ancient Italy. While the art of dentistry is almost as ancient as mankind (see below), many earlier civilizations such as the Egyptians had religious taboos about replacing parts of the body, so they hadn't developed the art of making false teeth. The first clear evidence we have of dentures comes from the Etruscans, in about 750 BC. They used human teeth or carved ox teeth, which were soldered into bridgework made of a band of gold. While some cultures have been ashamed of false teeth, the Etruscans seem to have used them as a status symbol – the gold band was designed in such a way as to be fully visible, while a pin through the band held the replacement teeth in place.

The Mayans of Central America developed their own twist on false teeth in about AD 600 – they are arguably the first culture to develop cosmetic dentistry. They regularly blinged up teeth by drilling tiny holes in them (using a rope drill and powdered quartz in water as an abrasive), and attaching miniature jewels into the holes using a natural resin glue. Modern museums contain some extraordinary examples of surviving skulls with such decoration. They also used pointed tools to carve patterns of notches and grooves into teeth. In addition, a jaw from the period has had three teeth replaced with decorative pieces of shell. Whether this was for the purposes of intimidation, status or display is unclear, but the effects of such tooth modification would certainly have been striking.

The final step to creating full dentures only came in the sixteenth century in Japan. Beeswax was used to create a mould, then the set of teeth was meticulously carved by craftsmen to match the model. The dentures were originally made of wood, but over the years they also used human teeth, ivory or pagodite, using advanced methods of adhesion to hold the entire structure in place in the mouth.

# Anaesthetics

## First used: Sumer
## Date: Fourth century BC

Historical evidence of surgical procedures suggests that some kinds of herbal anaesthetic were probably used in various parts of the world as long as tens of thousands of years ago. We have written evidence of a variety of substances being used from the earliest recorded times in history. For instance, alcohol was used to dull pain in ancient Mesopotamia, and it is thought that the Sumerians cultivated the opium poppy as long ago as the fourth millennium BC. The Sumerian goddess Nidaba was often depicted with poppies growing out of her shoulders, while the world's earliest pharmacopoeia (a text which can be used to identify medicines) is a Sumerian tablet from the third millennium BC – it talks about the opium poppy, which was known as *hul gil* (plant of joy).

By the time of the ancient Egyptians, mandrake root extracts were also commonly used as an anaesthetic during surgery. The opium poppy may also have been used, as it was over time in India and China – where cannabis incense and aconitum (also commonly known as monkshood or wolf's bane) were also known to have anaesthetic properties.

# A Very Brief History of Dentistry

The earliest (disputed) evidence of tool use for dentistry comes from Neanderthal culture.

- The first confirmed instance comes from an infected tooth found in Italy from 14,000 BC, which had been cleaned with flint tools.

- Using handheld drills to remove or drain infections is first seen in the Indus Valley civilization of 7000 BC.

- The earliest filling was found in a jaw in Slovenia, dated at about 4500 BC, and made of beeswax.

- It was a common belief from this period up to the Middle Ages that toothache was caused by toothworms, whose removal or destruction would alleviate the pain.

- Toothpicks were used from at least 3000 BC (having been found in ancient Sumerian sites).

- In ancient Egypt in about 2650 BC, the skills of Hesi-Re, 'greatest of all physicians and dentists', included wiring loose teeth in place using gold wire.

- The Code of Hammurabi (1800 BC) mentions dental extractions as a form of punishment (the rule of 'an eye for an eye and a tooth for a tooth' is first mentioned in this book of law).

- Pliny the Elder (AD 23–79) recorded tooth-related superstitions including tying a frog to your jaw, or using eardrops made of olive oil in which earthworms had been boiled.

- About 2,000 years ago, the Romans were using sticks as toothbrushes. Unfortunately they were using them with pastes that contained abrasives, such as emery, resulting in teeth being worn down to the roots.

- By the medieval period, dentists in China had developed a filling made of mercury, silver and tin (while Western dentistry was using more squalid fillings such as candle wax, lead and even 'raven's dung').

- Bristled toothbrushes were invented in China about AD 1000 – they were used with a tooth powder made out of soap beans.

# False Limbs

## First Invented: Greece
## Date: 479 BC

The use of false limbs to replace those that had been lost in accidents or warfare goes back to the first millennium BC, when it developed independently in a few different cultures including India and Egypt. The first documented case comes from the writings of Herodotus, the 'Father of History'. He describes the case of Hegesistratus, a Greek diviner who worked for the Persian Army as they were attempting to invade Greece. Hegesistratus had been captured by the Spartans, locked up with one foot in the stocks, and warned he was to be tortured. He managed to escape by cutting off part of his own foot and fleeing in great pain. Such was his hatred of the Spartans, he had a wooden replacement made, purely so that he could take part in the ensuing Battle of Plataea in 479 BC. Unfortunately his determination and animosity were not enough to save his life. He was captured by the Spartans once more, and put to death.

One remarkably fine example of a false limb comes from a tomb at Capua in Italy, dated to about 300 BC. A wooden lower leg, stabilized with thin bronze sheeting and an iron pin, it has a crafted concave top to hold the stump of the thigh. Also in the Roman era, Pliny records that Sergius Silus, a veteran of the Punic Wars, had been wounded in battle twenty-three times and had an iron hand built to replace the one he lost while fighting.

## Early Tattoos

The art of body modification, and tattooing in particular, comes with a long heritage. We know from a combination of preserved skin, ancient art and the discovery of ancient tattooing tools, that tattoos have been with us since the Upper Paleolithic period (50,000–10,000 years ago). The first direct evidence of tattooing that we have is the mummified skin of Ötzi the Iceman, from about 3200 BC. Other tattooed mummies have also been found around the world, for instance in Greenland, Mongolia, the Philippines and the Andes. One well-known example is the tattooed mummy of Amunet, Priestess of the Goddess Hathor, from ancient Egypt (c. 2134–1991 BC).

## CHAPTER 6

# SCIENTIFIC
# ADVANCES

# Magnetism

## First Discovered: Ancient Greece
## Date: 600 BC

The story is that magnetism was discovered by a Greek shepherd called Magnes, who lived in Magnesia (the Aegean coast of modern-day Turkey). He found that the buckle of his sandal was being attracted to a particular rock and, after experimenting, realized this also happened to other metal objects. After digging into the ground, he found that lodestones, which contain magnetite (a naturally magnetic material), were the cause.

Magnets were given the name of *magnetis lithos* (which means the rock of Magnesia). Their use gradually spread around the world – for magic displays, and for recreational and practical purposes. Pliny the Elder (AD 23–79) wrote about a hill near the river Indus containing a stone that magically attracted iron. It was a common superstition among sailors that an entire magnetite island might exist, and that the magnetic effect on the iron nails of passing ships would lead to shipwrecks. Magnetite was even credited with the ability to heal the sick and repel evil spirits.

It took a few more centuries before the navigational application of the magnet was discovered. The Chinese were using magnets for wet compasses (see page 90) by the eleventh century AD, and between 1405 and 1433 the famous Admiral Zheng He journeyed across seven oceans while using a magnetic needle compass. The earliest European mentions of magnets as a navigation device are in the twelfth century AD. The word commonly used in Europe was lodestone (*leading stone* in Anglo-Saxon or *leider-stein* in Icelandic).

Today we know that the Earth itself is a giant magnet, with two magnetic poles. This was first confirmed by William Gilbert, an English scientist, in 1600, while the connection between electricity and magnetism was first proven by the Dutch scientist Hans Christian Oersted in 1820.

# Pigment

## First used: Zambia
## Date: 350,000 years ago

The pigments that were the traditional base ingredients of paints were gradually discovered by humankind over many millennia. As recently as the fifteenth century, it was still extremely difficult to find lapis lazuli, the blue mineral that was the main ingredient in ultramarine paint. The great painter Jan van Eyck would only include blue in a portrait if the client paid extra, and when he did it would be in small quantities, unblended with other colours, and highlighted to emphasize its use. This is also the reason why Jesus's mother Mary was traditionally depicted swathed in blue, to emphasize her high status.

Lapis lazuli was in use at least 6,500 years ago, and was highly valued in the early civilizations of Mesopotamia, Egypt and China. There are still lapis lazuli mines in Badakshan, a remote, mountainous region of Afghanistan – they have been continuously in existence since at least 700 BC, when they were part of a country known as Bactria.

Purple was also a traditionally expensive colour. Tyrian Purple was a pigment made from the mucus of the Murex snail. It was in use as a fabric dye in about 1200 BC. Green paint, which was in use from prehistory, was also relatively rare, being primarily derived from the clay minerals of celadonite and glauconite.

However, there were some colours that were in plentiful supply from the early history of humankind. Carbon black can be made from charred bones or wood; yellow and red come from iron oxides; brown, in a variety of shades, can be made from iron and manganese oxides; and white can be made either from chalk or from animal bones. All of these pigments could be used to colour human skin or the walls of caves, either by directly applying them or by mixing them with animal fat and thus making a primitive paint mixture. The earliest evidence we have of painting comes from at least 350,000 years ago – equipment for grinding paint and pigments was found by archaeologists in a cave at Twin Rivers, which is close to Lusaka in modern-day Zambia.

# Algebra

## First Invented: Persia
## Date: AD 830

Given that ancient Greek and Indian mathematicians were able to calculate complex problems such as the value of pi and the square root of two, to high degrees of accuracy, it is hard to imagine that they weren't using algebra – a method of mathematical problem-solving that we now take for granted. However, the Greeks and Indians were using far more cumbersome techniques, and it was not until the ninth century that algebra as we now recognize it was invented by the Persian mathematician Muhammad ibn Mūsā al-Khwārizmī. His treatise *Kitāb al-muḫtaṣar fī ḥisāb al-ǧabr wa-l-muqābala* was translated into English as The Compendious Book on Calculation by Completion and Balancing. The word *al-ǧabr* or 'completion' referred to the practice of moving terms from one side of an equation to the other, while *muqābala* referred to the practice of reducing equations by dividing both sides by the same amount. The spread of his ideas around the Near East, China and Europe led to a gradual revolution in the way that mathematics was understood and practised.

# The Atom

## First Described: Ancient Greece
## Date: Fifth century BC

Our modern concept of the atom was inspired by John Dalton, the early nineteenth-century English scientist who suggested that atoms were the fundamental, indivisible building blocks of everything in the universe. His work was essentially accurate, although we do now know that atoms are divisible into smaller subatomic particles such as protons, neutrons and quarks.

However, Dalton's work was merely the latest version of an ancient philosophy. Both Indian and Greek thinkers from over two millennia ago had suggested that everything in reality was made up of indivisible particles. Pre-Socratic philosophers (meaning those who came before Socrates) in Greece were fascinated by the nature of reality. Some of the most prominent were Heraclitus (who believed that all of reality is change), Parmenides (who argued that change itself was an illusion), and Democritus, who, following in the footsteps of his teacher Leucippus, tried to reconcile these two systems of thought by arguing that reality was made up of tiny indivisible particles which he named 'atoms'.

Meanwhile, in the same period, the Jain philosophy of India envisioned the world as being made up of atoms, and set out a complex deterministic theory of the ways that these atoms interacted to form the substances which make up the universe. Both of these ancient versions of atoms differed somewhat from our modern theory, but it is notable that the fundamental concepts of Democritus and the Jain philosophers were fairly accurate even though they were not correct in every detail.

# Counting

## First Invented: Africa
## Date: 40–50,000 years ago

It is most likely that counting started with the fingers, which allowed a count of ten. This would have been enough for most purposes when combined with (for instance) an open arm gesture for 'many'. The first 'tallies' come from at least 40,000 years ago. A baboon fibula found in the Lebombo Mountains in southern Africa is marked with twenty-nine notches – one interpretation is that it was being used to record the lunar cycle, although as it is broken this seems an overconfident suggestion. Many other bones and other items have been found dating to subsequent millennia with similar tallies.

The first written system for record keeping that we know of comes from the pre-Sumerian period in modern-day Iran. Early tokens were made of clay with a symbol representing one sheep, ten sheep, one goat, ten goats and so on. These clay tokens were strung together and baked into a clay envelope, which was marked on the outside to prevent tampering (it was to be kept for the purpose of checking that the correct number of livestock had been returned or exchanged at a later date).

This was a precursor to the Sumerian civilization, which used a variety of different methods of recording numbers, and seems to have been the birthplace of arithmetic. Scribes responsible for keeping records of quantities of grain on a daily basis started to find ways to practise and improve their skills and speed in adjusting the quantities that they were writing down.

# Zhang's Seismoscope

## First used: China
## Date: Second century AD

A seismometer (or seismograph) is a scientific instrument that measures distant earthquakes and volcanic activity through the tiny movements of the ground that they cause. Remarkably, the first seismometer was invented nearly 2,000 years ago in China. It is known as Zhang Heng's seismoscope.

Its inventor Zhang Heng (see page 167) thought that the main cause of earthquakes was chaotic air motion, theorizing that:

> ... as long as [air] is not stirred, but lurks in a vacant space, it reposes innocently, giving no trouble to objects around it. But any cause coming upon it from without rouses it, or compresses it, and drives it into a narrow space ... and when opportunity of escape is cut off, then 'With deep murmur of the Mountain it roars around the barriers', which after long battering it dislodges and tosses on high, growing more fierce the stronger the obstacle ...

The seismoscope: a modern recreation of Zhang Heng's apparatus for detecting earthquakes.

In Zhang's device the earth tremors made a bronze ball fall out of any one of eight tubes (in the shape of dragons' heads). The ball then fell into the mouth of a metal toad, whose position indicated the orientation of the seismic wave. It isn't fully known how the device worked. The eight mobile arms definitely raised a catch via a crank, and a lever, which released the ball. Apparently the device also included a pendulum hung from a bar. This suggests that the driving force was inertia – a small movement in the pendulum perhaps triggered the motion that was transformed into a slight force on the correct lever.

However, there are no clear historical documents or remaining examples, so modern attempts at reconstructions involve a considerable degree of speculation and interpretation of the few mentions that were made of the device in contemporary texts. Nonetheless, it seems clear that Zhang's seismoscope did use similar technology to early modern seismographs. After the death of Zhang, it was not until 1783 that a simple seismograph was deployed by an Italian scientist called Schiantarelli, who used it to measure a major earthquake in Calabria.

## Zhang Heng, the Chinese Da Vinci

It's worth taking a moment to celebrate the life of Zhang Heng (AD 78–139), whose expertise across a wide range of fields – including maths, science, engineering, cartography, art and poetry – have led to him being described as the Leonardo da Vinci of ancient China. He was initially a minor civil servant, but rose to become the Chief Astronomer and Palace Attendant at the imperial court. As well as his seismoscope, he invented a water-powered astrolabe (a three-dimensional model of the solar system), gave an improved estimate for the number pi, and catalogued over 2,500 stars. He also gave an advanced description of the moon, its 'dark side', and how lunar and solar eclipses proved that the moon must be a spherical object. And if all that wasn't enough for one individual, he was also a renowned poet, whose work was still being studied years after his death.

# The Star Chart

## First Invented: Europe
## 32,500 years ago

As well as trying to understand the movement of the sun and the moon (see page 12), early mankind was fascinated by the stars, which were often identified with deities. The earliest known chart of the stars is a carved mammoth tusk found in Germany, with a carving that closely resembles the constellation of Orion (although this is disputed). A wall drawing in the Lascaux caves in France (which could be dated any time from 33,000 to 10,000 years ago) has a clear representation of the Pleiades cluster. In the same caves, there is an image of a bison, a man with a bird's head and the head of an actual bird. It has been suggested that this is also an astrological drawing, representing the summer triangle (a triangle of some of the sky's brighter stars – Altair, Deneb and Vega). More detailed star catalogues started to be used from the Babylonian and ancient Egyptian periods, from the second millennium BC onwards, but it is likely that these reflected a body of knowledge that had been gradually built up over previous millennia. The earliest known chart of the stars is a carved mammoth tusk found in Germany: it has a carving which

The outlines on this prehistoric cave painting form a map of the sky with the eyes of the bull, birdman and bird representing the three prominent stars Altair, Deneb and Vega.

closely resembles the constellation of Orion (although this is disputed). A wall drawing in the Lascaux caves in France (which could be dated any time from 33,000 to 10,000 years ago) has a clear representation of the Pleiades cluster. In the same caves, there is an image of a bison, a man with a bird's head and the head of an actual bird, representing the three prominent stars Vega, Deneb and Altair. Together, these are popularly known as the Summer Triangle and are among the brightest objects that can be picked out high overhead during the northern summer.

# Mapmaking

At the same time as humankind started to depict the layout of the heavens, they also started to do the same for their more immediate surroundings. There are some cave paintings and rock carvings that show landscape elements such as mountains and rivers – for instance, an image showing the main features of the landscape around Pavlov in the Czech Republic has been dated to 25,000 BC, and a similar type of proto-map has been found on a 16,000-year-old chunk of sandstone in Navarre in Spain. A fragment of mammoth tusk found in the Ukraine shows the plan of a stream with a row of houses next to a river.

By the time of the Babylonian civilization, mapmaking had advanced to the point where advanced surveying techniques were being used to create accurate maps which recorded, amongst other details, the ownership of plots of land. The first known map of the world, from about 600 BC, is also Babylonian – it shows the world as a giant circle surrounded by water, which was the religious model they believed in at the time.

# Pesticides

From the earliest days of agriculture, about 10,000 years ago, humankind has faced the problem of how to control pests that attack crops. The first records we have of pesticide use come from the ancient Sumerians, who used sulphur to address insect infestations. The Chinese used combinations of mercury and arsenic compounds for protection against body lice. By the time of the early Romans and Greeks, we have an increasing range of evidence of different methods that were used. Some of these were magical or folk myths, but there were also some methods that we would recognize today. Smoke, from burning straw or animal matter, was used to control mildew, blight and insects. Sticky substances such as tar were used to entrap insects crawling up stems and trunks. Salt was used to attack weeds, while plant material such as pyrethrum daisies was powdered and used to protect grain stores. And there is a multitude of other records of protective seed coatings, herbicides and rodenticides that were made from chemicals and minerals that could be prepared from native soils, vegetation and wildlife, many of which appear to have been in use for centuries before the classical period.

# The Rocket

## First Invented: China
## Date: AD 1232

The technology that would eventually take humankind to the moon and beyond originated in ancient history – the rocket was an offshoot of the development of gunpowder. It is possible that the first rockets were developed in the tenth century AD. Fire arrows, in which gunpowder was attached to arrows, were in use from 904, and it was this weapon that was eventually refined into the earliest rocket. The author Liang Jieming credited two generals named Yue Yifang and Feng Jisheng with the invention of a version of the fire arrow that converted gunpowder tubes into propellant in AD 969. However, more reliable evidence for the use of rockets comes from a few centuries later. There are reports of 'iron pots' whose explosions could be heard 25 km (16 miles) away, while rockets were used in a Song Dynasty navy exercise in 1245. While fireworks had been a popular element of festivities in China at least since the invention of gunpowder, the first mention of a rocket-propelled firework comes from a feast held by the Emperor Lizong of Song in honour of his mother, the Empress Dowager Gong Sheng – who was reportedly startled by the explosion of the device.

# The Camera Obscura

## First Described: China
## Date: Fourth century BC

The camera obscura (or pinhole camera) was the earliest way that people could project images onto a surface. The Latin name means 'dark room'. The method works best when a pinhole is cut in a surface which projects the image from brighter surroundings onto the opposite surface of a darkened chamber. The image will be inverted and reversed as the rays of light travel through the hole and cross over to the opposite side of the surface projection.

Some have suggested that this technique was known as long ago as the Stone Age. Paleolithic cave paintings show distortions in the shapes of animals, which could possibly have happened because the artist in the cave was tracing a slightly unstable projection on the cave wall. Small holes in some Neolithic structures might also have been inspired by the religious desire to project the sun's image.

At some point in the first millennium BC, Chinese sundials started to use a small hole in the gnomon (the part of the sundial that creates the shadow) to project a pinhole image of the sun, which gave a more accurate measure of the time of day. However, it is not until the fourth century

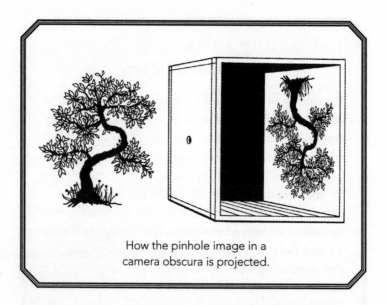

How the pinhole image in a
camera obscura is projected.

BC that we have a clear reference to the camera obscura, in the writings of a philosopher called Mozi. He described the way an inverted image appears in a 'collecting-point' or 'treasure house'.

It is thought that the ancient Greeks used pinhole projections to study solar eclipses and other phenomena. For instance, Aristotle comments on the conundrum that light from the sun travelling through a square hole nonetheless projects a circular image of the sun. But other than that, there are no clear Western descriptions of the camera obscura until the eleventh century, so this is probably another case in which the Chinese were a long way ahead of the West.

# The Number Zero

The concept of a number zero is an indispensable part of calculus, algebra, computation and our ability to clearly express large numbers. However, using zero as an actual number is a relatively recent idea in mathematical history. At some point in Babylonian and Sumerian history during the first three millennia BC, the idea of a positional number started to be used in their numerical system – meaning that the value of a number depended on its position (in the same way as 1 has a large value when it is the first digit of a two digit number such as 12). However, rather than writing a number in an empty position (as we would in the number 205), they left a space or marked the space with two dashes.

It was not until about the fifth century AD that Indian mathematicians started to treat zero as a real number. This meant that, in their system, the placeholder could be marked with a zero. It also made it much, much easier to express and imagine large numbers. Imagining a number like a million using Greek and Roman numerals was extremely cumbersome, whereas the Indians could now simply write a one followed by six zeros. This revolutionized the way that calculations were performed, and laid the groundwork for the arithmetic we all now learn in school.

# Refraction and Optical Devices

First Described: Persia
Date: Tenth century

One early kind of eyeglass was the flattened walrus ivory that prehistoric Inuits wore in front of their eyes for protection from the sun, creating makeshift sunglasses. But the true history of eyeglasses starts with the invention of glass-cutting and blowing (see page 86), when people noticed that if you looked at the world through curved glass it distorted your vision. There are ancient examples of what appear to be lenses, although their ability to magnify is limited and it may be that they were burning glasses, which started a fire by focusing the power of the sun. The Greek dramatist Aristophanes mentions 'the beautiful, transparent stone with which they light fires' in *The Clouds* (424 BC). Pliny the Elder (AD 23–79) suggests that glass balls filled with water are able to set clothes on fire when the sun shines through them. By the time of the Romans, the idea of a lens as a magnifier was also relatively well known. Emperor Nero held a polished emerald in front of his eyes so he could watch the gladiators fighting more clearly, while Seneca, his tutor, is said to have read books through a large glass bowl that magnified their contents.

The scientific breakthrough in our understanding of this notion came in AD 984, when the Arabian scientist Ibn Sahl wrote the treatise *On Burning Mirrors and Lenses*, describing the way in which curved mirrors and lenses bend and focus rays of light. This text included a statement of the law of refraction, which is now often called Snell's Law (when it should really carry the name of Ibn Sahl).

Ibn Sahl showed how different lens shapes could focus light without distortion, as well as discussing how ellipsoidal mirrors work. In the following century, 'reading stones' were being made in Venice, comprising a flat-bottomed convex sphere that could be laid over a book to create a functioning magnifying glass, although not one that was tailored for individual use. (In the twelfth century, Chinese judges sported crystal sunglasses made of smoky quartz for a different purpose – to conceal the expression on their faces from observers.)

Individually manufactured corrective glasses first developed in Europe in the late thirteenth century. The first visual evidence we have of this style of spectacles comes from paintings by Tommaso da Modena in the mid-fourteenth century, which show monks wearing both monocles and pince-nez (from the French for 'pinch nose') eyeglasses while they worked on manuscripts.

Optical devices took another leap forwards at the start of the seventeenth century. The Dutch eyeglass maker

Hans Lippershey filed a patent for a telescope in 1608; Galilei Galileo opportunistically tweaked the design the following year and used it to study the night skies (which means he is sometimes credited as being its creator). By 1611, Johannes Kepler had outlined ways to use multiple lenses to make more powerful telescopes, which would be used by astronomers to make extraordinarily detailed observations of our nearest celestial neighbours, such as the moon and Mars.

# A Brief History of Long-Distance Communication

You could post a letter from one place to another in at least the second millennium BC – the Assyrian civilization had a particularly efficient system by which letters, messages and payments could be transferred over significant distances.

- The first pigeon-post systems also date from about 2000 BC. In the Sumerian civilization it was known that pigeons would return to their home over great distances, and this was exploited to carry messages.

- Postal systems operated in the ancient Egyptian empire and in China by 1000 BC.

- In the Persian Empire in the sixth century BC, letters could be carried over 1,600 miles in just nine days by a relay of horse riders similar to the more recent Pony Express.

- The first long-distance takeaway food delivery was organized for Aziz, the caliph of North Africa, in the tenth century AD. His vizier had six hundred pigeons sent from the Lebanon to Cairo, each carrying a parcel containing one of his favourite cherries.

- In the fifth century BC, the Greek dramatist Aeschylus described the use of beacons to send news of the capture of Troy five centuries earlier.

- Aeneas Tacticus, in the fourth century BC, developed a method of using torches to send coded messages over a distance.

- In the second century BC, the Greek historian Polybius wrote of a more complex system of semaphore which also used one of the earliest codes or ciphers – two torches were used to indicate positions on a grid of letters that the sender and receiver had previously agreed.

- The Romans would use a variant of Polybius's method to send messages over hundreds of miles between a chain of signalling stations of forts around their empire.

- The smoke signals used by North American Indians operated on a similar basis to semaphore and fire beacons. They were in use when Columbus first arrived in America, but had probably been a longstanding tradition going back over the centuries.

# AFTERWORD

One of the fascinating things about exploring the world of ancient inventions is the perspective it gives you on the rise and fall of civilizations. Consider an Egyptian in the second millennium BC, or an Athenian in the second century BC, or an Italian living through the Renaissance. At any of these moments, as well as countless other times through history, there would have been a perception that human history was, on the whole, a process of moving forwards and becoming ever wiser, more scientific and more enlightened. The newest inventions would have been sources of wonder, and people would have felt that the future would hold more of the same – more progress, advances in technology and science and a gradual improvement in the human condition.

However, science and technology can be lost as well as found. Some of the inventions mentioned in this book may never be reconstructed or rediscovered. History has its many dark ages and cataclysms during which knowledge is lost and the human condition gets worse rather than better. So how comfortable should we feel today that the technology we take for granted will still be here in 1,000 years or more? The presumption tends to be that by then we will be out in space, reliant on advanced energy systems, and understanding ever more of the cosmos and subatomic physics. We might even have our jetpacks and flying cars – we've been waiting long enough after all.

But perhaps that isn't how things will be. Perhaps there will be a human civilization that has evolved into something unrecognizable to us, in which there remain only a few incredulous rumours about how mankind used to be able to fly to the moon and beyond, send information instantaneously around the world, cure all kinds of diseases, send aircraft and bombs across the oceans, and understand what matter is made of.

Let's hope that the utopian version is the real story and the dystopian one never comes to fruition. But in the meantime, take a look around and appreciate that every little bit of technology that you use, from the biro pen, to the paper notebook, to the light switch to the printing press that printed this book, was once just a pipe dream. The world we live in is amazing, and it's because someone invented all of these things, either in the ancient world or more recently. And in the future the one thing we can rely on is that human ingenuity will continue to transform the world for better or for worse.

# PICTURE CREDITS

Page 15: Drawn from a painting in the caves of Cueva de la Arana by Achillea / GNU General Public Licence version 2.

Page 17: Su Song clock tower illustration by Aubrey Smith.

Page 23: Carole Raddato / Flickr / CC BY-SA 2.0

Page 27: Gift of Theodore M. Davis, 1907 (07.226.1), Theodore M. Davis Collection, Bequest of Theodore M. Davis (30.8.54), Metropolitan Museum of Art, New York.

Page 29: Early distillation process illustration by Aubrey Smith.

Page 38: Rogers Fund, 1936, Metropolitan Museum of Art, New York.

Page 46: From *Science and Civilisation in China: Volume 5, Part 1, Paper and Printing* by Joseph Needham, Caves Books Ltd., Taipei 1986.

Page 49: *An Old Sufi Laments His Lost Youth*, ink and pigments on laid paper, 1597–1598 (Mughal), W.624.35A, acquired by Henry Walters, The Walters Museum of Art, Baltimore, USA.

Page 61: Queen Nefertari playing chess: fresco from the tomb of Nefertari, Thebes, 13th century B.C. / The Yorck Project, *10,000 Meisterwerke der Malerei,* DIRECTMEDIA Publishing GmBH.

Page 65: Illustration from Knight's American Mechanical Dictionary, Houghton, Mifflin and Company, Boston, 1882.

Page 69: Clipart.com.

Page 75: Science and Society Picture Library / Getty Images.

Page 79: Matyas Rehak / Shutterstock.

Page 81: Eric Gaba / CC SA-BY 4.0

Page 82: Archimedes screw illustration by Aubrey Smith.

Page 87: MatthiasKabel / CC SA-BY 3.0

Page 93: Windmills of Nashtifan, Khaf, Khorasan Razavi, Iran; photo by Ahmad Hassanzadeh / CC SA-BY 4.0

Page 98: Bequest of Richard B. Seager, 1926 / Metropolitan Museum of Art, New York.

Page 99: Purchased with the assistance of the National Art Collections Fund / British Museum (M&ME 1958.12-2.1); photo by Marie-Lan Nguyen / CC BY-SA 2.5

Page 106: DerHexer / Wikimedia Commons / CC SA-BY 4.0

Page 108: The Nimrud lens or Layard lens, British Museum (BM number 90959) ; photo by Geni / CC SA-BY 4.0

Page 113: Mogi / CC SA-BY 3.0

Page 119: Archimedes' heat ray illustration by Aubrey Smith.

Page 121: Reconstructions of ancient mechanical artillery, Saalburg Museum, Hesse, Germany; photo by SBA73 / Flickr / CC SA-BY

Page 125: Illustration from *Discorso della Religione Antica de Romani*, 1569.

Page 127: Illustration from *The Crossbow* by Sir Ralph Payne-Gallwey, Longmans, Green and Co., London 1903.

Page 166: Mary Evans / Natural History Museum.

Page 169: Universal History Archive / UIG / Getty Images

Page 174: Camera obscura illustration by Aubrey Smith.

# ACKNOWLEDGEMENTS

I'd like to thank everyone who read the first draft of this book and gave me helpful feedback – you all know who you are. Thanks to everyone at Michael O'Mara Books, especially Louise Dixon for commissioning this book and Gabby Nemeth for her care and attention to detail. And thanks to my daughter and, above all, my wife for putting up with me regaling them with fascinating facts about history.

# INDEX